Human Papillomaviruses

PERSPECTIVES IN MEDICAL VIROLOGY

Volume 8

Series Editors

A.J. Zuckerman

Royal Free and University College Medical School
University College London
London, UK

I.K. Mushahwar

Abbott Laboratories
Viral Discovery Group
North Chicago, IL, USA

Human Papillomaviruses

Editor

D.J. McCance

Department of Microbiology & Immunology
and the Cancer Center
University of Rochester
Rochester, New York 14642
USA

2002
ELSEVIER
Amsterdam – Boston – London – New York – Oxford – Paris – San Diego –
San Francisco – Singapore – Sydney – Tokyo

ELSEVIER SCIENCE B.V.
Sara Burgerhartstraat 25
P.O. Box 211, 1000 AE Amsterdam, The Netherlands

First edition 2002

Library of Congress Cataloging in Publication Data
A catalog record from the Library of Congress has been applied for.

British Library Cataloguing in Publication Data
A catalogue record from the British Library has been applied for.

ISBN: 0-444-50626-8
ISSN: 0168-7069

⊚ The paper used in this publication meets the requirements of ANSI/NISO Z39.48-1992 (Permanence of Paper).

Transferred to digital print, 2008
Printed and bound by CPI Antony Rowe, Eastbourne

Contents

Contents

Preface

Human papillomaviruses are the cause of benign, premalignant and malignant lesions of stratified epithelia. While benign warts were a recognized clinical lesion in Roman times, it was not until the 1970s that papillomavirus infection was associated with certain cytological properties of premalignant disease of the cervix. In the last two decades the evidence that papillomaviruses cause various epithelial cancers has been building, and it is now clear these viruses are the causative agent. However, while papillomaviruses cause benign and premalignant lesions, there are unknown factors that are necessary in combination with the virus for progression to malignant disease.

This volume looks at the epidemiological evidence for association of HPV infection and disease, at the biology of viral proteins and the immune response to infection, all of which contribute to the pathogenesis. The effort is concentrated on those viruses which cause premalignant and malignant disease, as most of our knowledge of the interactions of the virus with host cells comes from the study of the oncogenic viruses. For all papillomaviruses to replicate in the stratified epithelium they must stimulate keratinocytes, which are programmed for terminal differentiation, to re-enter the cell cycle and progress through G1 into the S-phase of the cell cycle. The S-phase is necessary so that the replicative machinery of the cell will be available for the virus to utilize. The early proteins of HPV, E6, E7 and E5, are involved in the G1 to S-phase progression and their role and functions are discussed. Two other early proteins, E1 and E2, are involved in the replication process itself and possibly in the control of transcription of the early genes. New functions of another, rather enigmatic early protein, E4, is also discussed. The pathogenesis of papillomaviruses is limited by the host's immune response and this is discussed with particular emphasis on the potential vaccines, which are being tested at present. The composition of a protective immune response is unknown, but there is evidence that both a T- and B-cell response is required with the latter being directed against conformational epitopes on the L1 and 2 capsid proteins. This volume will be of benefit to scientists, clinicians and students who want an up-to-date account of the progress in human papillomavirus biology.

Human Papillomaviruses
D.J. McCance (editor)
© 2002 Elsevier Science B.V. All rights reserved

Clinical aspects and epidemiology of HPV infections

Cosette Marie Wheeler
University of New Mexico Health Sciences Center, School of Medicine, Department of Molecular Genetics and Microbiology, Albuquerque, New Mexico 87131, USA

Overview

Human papillomaviruses (HPVs) are a ubiquitous group of viruses that infect the epithelium and are associated with a broad range of clinical manifestations. Numerous HPVs cause no apparent disease whereas others are involved with the development of benign conditions such as common hand and foot warts, rare disorders such as epidermodysplasia veruciformis (EV) and invasive cancers of various anatomic sites. The recent intense focus on HPVs is founded in the establishment of their causal relationship with invasive cervical cancer.

Infection by HPVs most often occurs in cells of the epithelium including the genitalia and elsewhere. In addition to extragenital HPV infections of the skin, infections of the mouth, esophagus, larynx, trachea and conjunctiva have been reported. Table 1 lists the different types of HPV [39] for which genomes have been cloned and their more common clinical manifestations.

The majority of HPV infections remain clinically inapparent; however microfoci of infection obviously contribute to their spread within populations. The specific modes of HPV transmission are not well understood. Infection of actively proliferating basal epithelium through microlesions is the presumed point of viral entry. The number of proliferating cells exposed and the infecting dose of virus may affect the outcome of the HPV infection, but again, little is known in this regard. Expression of early viral genes occurs within the proliferative and differentiating part of the infected epithelium whereas the late-gene expression is limited to the outer differentiated layers. Thus the full virus life-cycle and production of virion is tightly linked to and requires cell differentiation. Until recently, this requirement had prevented for many years the successful production of infectious HPV particles under culture conditions.

HPV infection of host cells can result in both permissive and persistent infections. Permissive infections are characterized by a complete virus life-cycle including virion production. Persistence of productive HPV infections appears to be common for durations of at least months although in some instances viral persistence may continue for years or decades. The extent to which long-term persistent HPV infections are productive has not been defined. It is, however, presumed that persistent infections of these longer durations can increase the risks for cancer outcomes. This risk association can be easily rationalized given the fact that like most small

Table 1

HPV types with cloned genomes and common clinical manifestations

HPV type	Clinical manifestations	Subfamily/ Genus	Accession #	Ref.
HPV 1	Plantar warts	E1	V01116	35
HPV 2	Common warts	A4	X55964	130)
HPV 3	Flat and juvenile warts	A2	X74462	95
HPV 4	Palmar and plantar warts	B2	X70827	72
HPV 5	EV lesions and carcinomas	B1	M17463	191
HPV 6	Genital warts, CIN, VIN	A10	X00203	40
HPV 7	Butcher warts	A8	X74463	131
HPV 8	EV lesions and carcinomas	B1	M12737	139
HPV 9	EV lesions	B1	X74464	96
HPV 10	Flat warts	A2	X74465	95
HPV 11	Genital warts, CIN, laryngeal, nasal and conjunctival papillomas	A10	M14119	61
HPV 12	EV lesions	B1	X74466	95
HPV 13	Oral hyperplasia	A10	X62843	138
HPV 14	EV lesions and carcinomas	B1	X74467	94
HPV 15	EV lesions	B1	X74468	94
HPV 16	Warts, CIN, VaIN, carcinomas of cervix, penis, bronchus	A9	K02718	43
HPV 17	EV lesions and carcinomas	B1	X74469	94
HPV 18	Warts, CIN, VaIN, carcinomas of cervix and penis	A7	X05015	21
HPV 19	EV lesions	B1	X74470	94
HPV 20*	EV lesions and carcinomas	B1	U31778	59
HPV 21	EV lesions	B1	U31779	94
HPV 22	EV lesions	B1	U31780	94
HPV 23	EV lesions	B1	U31781	94
HPV 24	EV lesions	B1	U31782	94
HPV 25	EV lesions	B1	U74471	59
HPV 26	Cutaneous warts, CIN	A5	X74472	133
HPV 27	Cutaneous warts	A4	X73373	132
HPV 28	Cutaneous warts	A2	U31783	54
HPV 29	Cutaneous warts	A2	U31784	50
HPV 30	Laryngeal carcinoma, CIN	A6	X74474	41
HPV 31	CIN, carcinoma of cervix	A9	J04353	106
HPV 32	Oral hyperplasia	A1	X74475	13
HPV 33	CIN, carcinoma of cervix	A9	M12732	11
HPV 34*	CIN, Bowen's disease of skin	A11	X74476	86
HPV 35	CIN, carcinoma of cervix	A9	M74117	108
HPV 36	EV lesions, actinic keratosis	B1	U31785	85
HPV 37	Keratoacanthoma	B1	U31786	150
HPV 38	Malignant melanoma	B1	U31787	150
HPV 39	CIN, PIN, carcinoma of cervix	A7	M62849	12

continued

Table 1 (continuation)

HPV type	Clinical manifestations	Subfamily/ Genus	Accession #	Ref.
HPV 40	CIN, PIN	A8	X74478	37
HPV 41	Cutaneous warts	NA	X56147	69
HPV 42	Genital warts, CIN	A1	M73236	12
HPV 43	Genital warts, CIN	A8	M27022	107
HPV 44	Genital warts, CIN	A10	U31788	107
HPV 45	CIN and carcinoma of cervix	A7	X74479	124
HPV 47	EV lesions	B1	M32305	1
HPV 48	Squamous carcinoma of skin	B2	U31789	122
HPV 49	Cutaneous warts	B1	X74480	53
HPV 50	EV lesions	B2	U31790	52
HPV 51	CIN, carcinoma of cervix	A5	M62877	126
HPV 52	CIN, carcinoma of cervix	A9	X74481	159
HPV 53	Normal cervix, CIN	A6	X74482	58
HPV 54	Genital warts	A7	U37488	51
HPV 55	Bowenoid papulosis	A10	U31791	51
HPV 56	CIN, carcinoma of cervix	A6	X74483	41
HPV 57	CIN, cutaneous and nasal warts	A4	X55965	37
HPV 58	CIN, carcinoma of cervix	A9	D90400	116
HPV 59	CIN, VIN, carcinoma of cervix	A7	X77858	142
HPV 60	Cutaneous warts, epidermoid cyst	B2	U31792	115
HPV 61	Normal cervix, CIN, VaIN	A3	U31793	117
HPV 62	Normal cervix and CIN, VaIN	A3	U12499	117
HPV 63	Myrmecia wart	E1	X70828	44
HPV 65	Pigmented wart	B2	X70829	44
HPV 66	CIN and carcinoma of cervix	A6	U31794	170
HPV 67	CIN, VaIN, and carcinoma of cervix	A9	D21208	117
HPV 68	CIN and carcinoma of cervix	A7	X67161	104
HPV 69	CIN, VaIN, and carcinoma of cervix	A5	AB027020	164
HPV 70	Vulvar wart and CIN	A7	U21941	104
HPV 71	VaIN	A15	AB040456	164
HPV 72	CIN, oral warts	A3	X94164	176
HPV 73	CIN, oral warts and carcinoma	A11	X94165	176
HPV 74	VaIN	A10	U40822	105
HPV 75	Cutaenous warts	B1	Y15173	42
HPV 76	Cutaneous warts	B1	Y15174	42
HPV 77	Cutaneous warts, carcinoma of skin	A2	Y15175	42
HPV 78	NA (cutaneous)	A2	NA	NA
HPV 79	NA (genital)	A8	NA	NA
HPV 80	Normal skin	B1	Y15176	42
HPV 81	NA (genital)	A3	NA	NA
HPV 82	VaIN	A5	AB027021	88
HPV 83	Normal cervix	A3	AF151983	26
HPV 84	Normal cervix	A3	AF293960	171

continued

Table 1 (continuation)

HPV type	Clinical manifestations	Subfamily/ Genus	Accession #	Ref.
CandHPV 85	Genital wart	A7	AF131950	31
CandHPV 86	Genital, CIN	A3	NA	NA
CandHPV 87	Genital, CIN	A3	NA	NA

*Denotes an HPV type that has two separate types assigned to its identity but only one is listed here. HPV 46, although assigned, is considered the same as HPV 20. Similarly, HPV 64, although assigned, is considered the same as HPV 34. Ref. denotes the published article describing the cloning of the HPV type if available. When available, accession numbers (#) are provided for GenBank access to HPV genome or fragment sequence information. Cand = candidate HPV type from PCR, CIN = cervical intraepithelial neoplasia, EV = epidermodysplasia verruciformis, PIN = penile intraepithelial neoplasia, VaIN = vaginal intraepithelial neoplasia, VIN, vulvar intraepithelial neoplasia, NA = no available information or limited information. (Ref. [39], E.-M. de Villiers, personal communication

DNA viruses, HPVs disrupt the host cell regulatory machinery by harnessing it to propagate themselves. Thus, a simplistic view of long-term HPV persistence would support an increased probability of a malignant event within a disrupted cell environment. Because HPV-associated cancer outcomes are uncommon in comparison to the observed widespread nature of HPV infections, host and viral cofactors associated with HPV persistence are an important area of investigation that will be discussed later in the context of epidemiological findings.

It appears that HPVs along with other human pathogenic viruses such as herpesviruses, adenoviruses, and polyomaviruses developed prior to the emergence of humans. Papillomaviruses have been found in birds, reptiles and many mammals. Currently over 85 distinct HPV genotypes have been identified where complete genomic sequence is available. Accumulating data based on subgenomic sequences suggests that more than one hundred additional HPV genotypes exist. This extensive genomic heterogeneity is truly unique among DNA viruses and data support a remarkable ancient history of virus adaptations rather than a rapid acquisition of genome modifications. Epidemiological studies have demonstrated that gene sequences from various HPVs isolated from discreet parts of the world are remarkably conserved. This is true for both common and rare HPV types. HPV genotypes are now defined as having less than 90% identity in DNA sequence to any other reference genome and subtypes and variants of genotypes are defined as having greater than 90% and 98% identity, respectively. This identity can usually be defined by sequence comparisons limited to the L1 open reading frame (ORF) that encodes the major HPV capsid protein.

Methods for detection of HPV genomes have progressively developed over the past two decades. Initially cloned HPV genomes were used as probes in hybridization techniques such as Southern blot and dot blot hybridization. These methods for detecting HPV genomes yielded relatively specific results but were also insensitive. The advent of polymerase chain reaction (PCR) methods provided increased sensitivity and genotype-specific oligonucleotide probes enhanced the specificity of

HPV type-specific identification. Currently there are numerous broad-spectrum and type-specific PCR-based HPV detection methods that have been applied in elaborating the epidemiology of HPV infections. Extensive characterizations of populations using PCR-based HPV detection methods and more recently serologic assays have been primarily limited to studies of female populations and genital infections although these methods have facilitated a few studies of HPV infections in males and at extragenital sites. This review will, however, focus mainly on the epidemiology of genital HPV infections.

Epidemiology of HPV infections

Detection of genital HPV

An important aspect of elaborating the epidemiology of genital HPV infections has been the evolution of HPV sampling and detection methods. Estimates of HPV infection are very dependent on the populations sampled, the specimen collection methods and devices employed [63,137], the type of sample (i.e., fresh vs. archival samples) and laboratory approaches used for HPV DNA detection [22,65,153]. Sensitivity and specificity of the laboratory methods can vary when applying the same overall method such as PCR. If different primers, probes and protocols are used for the PCR, then the estimates are likely to differ. Furthermore, even the use of the same primers and probes with a protocol that varies slightly can result in differences in estimates. Because of this, direct comparison of HPV prevalence reported in various studies is difficult.

As mentioned earlier, a variety of PCR-based HPV DNA detection methods have been reported. These include type-specific HPV assays and broad-spectrum HPV typing assays such as the GP5+/6+ system [36,79], the MY09/MY11 system [9,114], a modified MY09/MY11 system designated PGMY09/PGMY11 [66] and the SPF-10 line probe assay [92]. None of these assays that amplify HPV DNA targets are approved for diagnostic use in the United States (U.S.). Currently only a signal amplified HPV DNA test, the Hybrid Capture II (HC2) assay is approved for diagnostic use. Recent data suggests that HPV testing by the HC2 assay is a viable option in the management of some low grade abnormal Pap smears designated as atypical squamous cells of unknown significance (ASCUS). The HC2 test was shown in a large randomized clinical trial to have greater sensitivity to detect cervical intraepithelial neoplasia (CIN3) or above and a sensitivity that was comparable to a single cytologic test indicating ASCUS or above [162]. In this same clinical trial, the high percentage of women (82.9%, 95% confidence interval (CI) = 79.7–85.7%) who were positive for HPV DNA by HC2 limited the potential of this test to direct the management of women with abnormal Pap smears designated as low-grade squamous intraepithelial lesions (LSIL) [172].

PCR-based data has demonstrated that sampling at a single time point results in an underestimate of HPV DNA prevalence [121,179]. This should be considered along with the fact that most reported risk factor associations have been based on single cross sectional HPV detection measurements. HPV DNA estimates in such studies represent a component of persistent infections along with newly acquired infections. The proportions of persistent and new infections cannot be estimated accurately and thus risk factors associated with new HPV infections that have been defined in cross sectional studies are of limited value. Longitudinal cohort studies particularly in populations that are susceptible to HPV can ascertain incident infections and overcome these obvious limitations.

Transmission of genital HPV

PCR-based molecular epidemiological data accumulated for more than a decade has demonstrated that detection of HPV DNA is strongly associated with sexual behavior including lifetime and recent number of sex partners. Prior to the generation of these molecular data, clinical evidence for sexual transmission of HPV infections was provided for both genital warts and for CIN. Oriel reported that 64% of partners of individuals with genital warts developed warts [129]. Similarly Barrasso [8] and Schneider [155] have reported HPV infections in male sexual partners of women with CIN.

The sharing of HPV types among sex partners has also been evaluated at the molecular level. In the study of Schneider and coworkers [155], 87% of male partners shared specific HPV genotypes detected in their partner's cervical specimen. In studies conducted by Ho [77] and Xi [185], analysis of HPV 16 variants demonstrated concordance of specific variants between sex partners although distinct variants were found among some couples. Detection of identical variants among sex partners provides some evidence for sexual transmission however the probability of detecting any given HPV 16 variant is also a function of its prevalence in the reference population. In studies conducted in U.S. populations, a single phylogenetic branch of HPV 16 variants would be found in over 70% of HPV 16 infected individuals [181]. Thus the likelihood of detecting these particular HPV 16 variants in any individual would be extremely high. It is probably important to note that the detection of identical HPV variants among sex partners does not provide any evidence for who was initially infected.

Regarding non-sexual transmission, there is some evidence for *in utero* infection [157,188], perinatal infection [160,169], auto- and hetero-inoculation through close non-sexual contact [70,127], and potentially indirect transmission via fomites [144]. Modes of HPV transmission among children remain controversial and the frequency of perinatal infections progressing to clinical lesions is unclear. Condylomata have been detected in the first week of life [169] and HPV DNA has been detected in both nasopharyngeal aspirates in newborns [157] and amniotic fluid [188]. In addition, laryngeal papillomatosis has been reported in infants [83,161]. In children, the overall prevalence of anogenital HPV is generally considered to be low. Determinants of HPV transmission are really unknown although levels of HPV DNA may play a role [87]. A recent article by Syrjanen and Puranen [168] provides a detailed review of HPV infections in children and the role of maternal transmission.

Genital HPV prevalence, incidence and global distribution

HPV prevalence provides a measure of the percentage of persons in a population who have new, persistent, or recurring HPV infection at a particular point in time. Prevalence can vary several fold, depending on the method of HPV detection and the demographic and sexual behavior characteristics of the group under study. HPV is detected in a large number of cytologically normal individuals and in most genital neoplasias and cancers. PCR-based point prevalence for genital HPV infection in women with cytologically normal Pap smears has ranged from about 1.5% in sexually inexperienced women to about 45% in sexually active women (147,180). Studies that have conducted repeated testing over time have demonstrated prevalences exceeding 50% in young women [179]. HPV 16 has been the most common HPV type detected among cytologically normal women and it is also the most common HPV type detected in cervical cancers worldwide [151]. An extensive review detailing existing PCR-based data has been published by Xi and Koutsky [186].

HPV infection is similarly common among men however studies in male populations are far more limited. In male sex partners of women attending an STD clinic, 63% of penile samples were HPV positive by PCR [7]. In healthy men aged 18 to 23 years, HPV DNA was detected by PCR in urethral specimens from 12% of 66 men with normal penile epithelium and in 26% of 39 men with aceto-white epithelium [84]. Lazcano-Ponce and coworkers [100] demonstrated that in urethral and coronal sulcus swab samples, HPV was not detected in men who reported not having engaged in sexual intercourse but was present in 43% of men who reported sexual activity. Case-control studies have been conducted to consider the potential role of the male factor in cervical cancer [18,123]. Twenty-six percent of husbands of 210 women with cervical cancer and 19% of husbands of 262 control women were positive for HPV DNA by PCR in Colombia whereas 18% of husbands of 183 women with cervical cancer and 4% of husbands of 171 control women were positive for HPV DNA in Spain.

HPV incidence provides a measure of newly acquired HPV infections in a population of persons during a specific time interval. Estimates of HPV incidence are generally limited to very defined study populations such as family planning clinics, sexually transmitted disease (STD) clinics, or university student populations. Ho and coworkers [76] reported a 14% annual incidence rate of subclinical HPV infection detected by PCR assays in college students. Woodman and coworkers [184] reported that in 1075 women who were cytologically normal and HPV negative at recruitment, the cumulative risk at 3 years of any HPV infection was 44% (95% CI 40–48); HPV 16 was the most common type. Population-based data based on HPV DNA results are essentially non-existent for incident HPV infection although clinical observations have been used to propose population-based incidence. The use of clinical measures to estimate HPV incidence are likely to represent underestimates since direct measurement of HPV DNA would be expected to result in greater detection sensitivity. Based on Pap smear cytology, an estimated crude annual HPV incidence of about 7% was reported for a cohort of women 22 years of age in Finland [167]. For genital warts, an incidence of 106 per 100,000 persons was reported in a population-based study of genital warts conducted in the United States [32]. Although data are extremely limited, HPV incidence is likely to be similar among women and men.

Data examining time trends for HPV infections are relatively limited and are confounded by the fact that awareness of HPV infections and their clinical manifestations increased significantly over the past 30 years. In addition, significant changes in diagnostic classifications occurred over this same time period. A U.S. survey of physicians reported that genital wart infections increased 4.5 fold during the years 1966 to 1984 [14]. Consistent with this observation, the incidence of genital warts reported in England and Wales doubled from 1971 to 1979 [5]. Increases in incident HPV infection might be expected to correlate with increasing incidence rates for other STDs. Increases in STDs have been observed when the proportion of individuals in the population who were young and sexually active increased. HPV seroprevalence studies have suggested an increase in HPV seropositivity across similar time periods. A population-based sample of pregnant women in Stockholm, Sweden between 1969 and 1989 found a 50% increase in HPV seroprevalence from 1969 to 1983 but stable seroprevalences during the 1980s [2]. The seroprevalence of herpes simplex type 2 (HSV-2) in these same samples showed similar trends [2], reflecting the increased rate of sexual activity in the population.

Genomes of HPV types and their variants are stable, since identical variants have been found in unrelated individuals residing in different countries who have no known contact with each other [30]. For HPV 16, five distinct phylogenetic branches have been reported. These branches have been designated E (European), As (Asian), AA (Asian American), Af-1 (African-1), and Af-2 (African-2) [30,190]. Studies conducted in numerous populations support the notion that representative variants from all of these five major HPV 16 lineages can be detected worldwide, although specific prevalences differ by geography [189]. Differences in HPV type and variant prevalences may be explained by founder effects and/or may reflect selection by the host population.

Determinants of genital HPV detection and persistence

Epidemiological studies in diverse populations that consider sexual, behavioral and demographic factors have generally concluded that detection of HPV decreases with age [10,27,38,119] and increases with number of sex partners both lifetime and recent. An extensive review of risk factors associated with detection of genital HPV infections was published by Xi and Koutsky [186]. Other sexual behaviors such as age at first sexual intercourse, years since first intercourse, frequency of sexual intercourse, sexual intercourse during menses, and anal intercourse have not been consistently associated with HPV DNA detection. Additional risk factors for detecting HPV may be population dependent and some are probably markers of sexual behavior. Studies examining the association of smoking and HPV detection have generally been negative. The relationship between oral contraceptive (OC) and sexual activity has made it difficult to determine the relationship between HPV detection and OC use. Several studies have reported positive associations between HPV detection and OC use although the majority of studies have not confirmed these findings. The association between HPV infection and reproductive history such as age at menarche, stage of menstrual cycle, age at first pregnancy, number of pregnancies and current pregnancy has also been inconsistent.

Few population-based studies have investigated the prevalence of type-specific infection for a broad spectrum of HPV types. Most studies have combined all genital HPV types detected into a single group for analytical purposes. Several studies have examined the determinants of high-risk and low-risk HPV types grouped separately [57,73,75,80,90,143,146]. Low-risk HPV types have been reported to be less associated with sex history and age than high-risk HPV types, although differences in associations have been reported. More recent investigations using highly sensitive and type-specific HPV DNA detection have demonstrated two peaks of increasing HPV prevalence. Of 9175 women in Guanacaste, Costa Rica, 3024 women were tested for more than 40 types of HPV with a PCR-based system [73]. Among women with normal cytology, HPV infections peaked first in women younger than 25 years, and then peaked again at age 55 years or older with predominantly low-risk and uncharacterized HPV types. Another population-based study was conducted in Mexico between 1996 and 1999 [99]. The sampling was based on an age-stratified random sample of 1,340 women with normal cytological diagnoses and 27 HPV types were distinguished. A first peak of 16.7% was observed in women under 25 years. HPV DNA prevalence declined to 3.7% in women 35–44 years and then increased progressively to 23% among women 65 years and older. This second peak of HPV infections in postmenopausal women demonstrated a clear predominance of cancer-associated or high-risk HPV types. The second peak of HPV prevalence in older women differed between these two studies in that one study reported a predominance of low-risk HPV types and the other reported a predominance of high-risk HPV types. The reason for these differences is not clear but further investigation are warranted to determine if the second peak of HPV prevalence in older women might reflect reactivation of latent HPV infections or newly acquired HPV infections following changes in immune and hormonal status. Certainly these data are intriguing given the possibilities as they relate to the natural history of cervical cancer outcomes. This second increase in HPV population prevalence in part overlaps the peak of cervical cancers in the population.

Studies on the persistence of cervical HPV DNA have been primarily limited to prevalent infections and time intervals between HPV DNA measurements have varied between studies. In addition, determinants of persistence have not been extensively examined. Despite these limitations, most studies have found that genital HPV infection is transient. Various investigations have demonstrated associations between HPV persistence and older age, types of HPV associated with cervical cancer, infection with multiple types of HPV and use of oral contraceptives [25,48, 75,76,103]. In one study the median duration of new HPV infections was 8 months (95% CI, 7 to 10 months) [76]. HPV type 16 has been shown in several studies to be the most persistent HPV type [25,45,103] followed by other high-risk HPV types.

Persistent detection of HPV 16 has demonstrated that the same dominant variant persists for months and sometimes years suggesting that reinfection by the same HPV type or of multiple variants is uncommon [185]. These data must be considered in the context of earlier discussions of expected risk for infection with particular HPV variants given population prevalences. Persistence has also been reported to increase with higher quantities of HPV DNA [25] although additional investigations in this area are needed. In a study conducted by Liaw and coworkers [101], persistence of concomitantly detected HPV was examined prospectively among 1124 cytologically normal women. Preexisting HPV 16 was generally associated with an increased risk for subsequent acquisition of other HPV types. HPV 16 did not affect the persistence of concomitant infections, regardless of type. This study suggests that prevention or removal of HPV 16 may be unlikely to promote the risk of infection with other HPV types. This has been a theoretical concern given prophylactic vaccination efforts that will be discussed later.

HPV and cancer

Epidemiological evidence has convincingly demonstrated that infection with high-risk HPVs is the greatest risk factor for cervical cancer [19,78]. Furthermore, the role of HPV in the development of CIN has been well established [15,93,152]. The relative risks for the association of HPV with CIN are commonly in the range of 20–70. This magnitude of risk is far greater than the association between smoking and lung cancer. Thus, HPV infection is considered a necessary but insufficient factor for malignant transformation.

Invasive cervical cancer occurs in approximately 400,000 women per year worldwide with an estimated 200,000 deaths per year [55,135,136]. The greatest burden of cervical cancer is in developing countries where it is often the most common female malignancy. Pap smear screening has reduced the incidence of cervical cancer in developing countries but in the U.S. alone, it has been estimated that over 5 billion dollars per year are expended to achieve a 75% reduction in cervical cancer [97].

An international study of invasive cervical cancers collected from 22 countries demonstrated that essentially all cervical cancers (99.7%) contained HPV [19,177]. Other studies reporting a proportion of HPV negative cervical cancer cases [56,102, 113,156] may have had specimens that were inadequate for testing, had extremely low copy numbers of HPV genomes or could have harbored integrated HPV forms that interfered with detection of targeted genomic segments. Numerous studies reporting HPV negative tumors have not included histological review to confirm the presence of tumor cells within the biopsy material and furthermore paraffin-embedded tissues have been used. The efficiency of amplification in paraffin-embedded tissues can be compromised [6,68] especially when the PCR targets are greater than a few hundred base pairs in length. In addition, the age of the specimen and variability in fixation methods can affect the amplification efficiency. The absence of HPV in a small proportion of cervical cancers has been reported as a poor prognostic factor for survival.

The most common HPV type detected in cervical cancers worldwide is HPV 16 followed by HPV 18 [19]. HPV 18 is more consistently associated with adeno-carcinomas of the cervix and less frequently with squamous cell carcinomas. Other types of HPVs found commonly in cervical cancers include HPV 31, 33, 45, 52, 58 and a few additional types at relatively low frequency. The prevalence of less common HPV types has probably been underestimated since large studies using broad spectrum HPV testing have not been reported.

Most studies have evaluated risk factors for squamous cell cervical cancer since it represents up to 90% of cases. The remainder of cases is mostly accounted for by adenocarcinomas and adenosquamous carcinomas. Both precursor CIN and invasive cervical cancers have been positively associated with number of lifetime sexual partners, age at first intercourse, and sexual behavior of the woman's male partners [24]. The increasing risk of infection with HPV as it relates to lifetime number of sexual partners and sexual behavior of a woman's male partner is understandable however, the association with age at first intercourse is somewhat less clear. It is possible that this association simply reflects total length of exposure to the causative agent, HPV.

Numerous additional risk factors for cervical cancer have been reported. Several studies have convincingly identified smoking as an independent risk factor for invasive cervical cancer [125,183]. The potential of direct carcinogenic effects from nicotine metabolites that can be found in cervical mucus has been speculated. Furthermore, smoking has been reported to down regulate the immune system, which might generally affect immune surveillance of HPV infections. Many studies have identified positive associations between cervical cancer and long-term OC use [140]. Hormone regulatory elements have been identified within HPV genomes and oestrogen stimulation has been shown to stimulate the expression of HPV 16 E6 and E7 in SiHa cervical cancer cells [120]. Regarding reproductive factors, several studies have found associations with multiparity and/or early age at first birth [134]. The hormonal changes and immunodepression associated with pregnancy could favor or enhance the transformation process and delivery, particularly of the first child could alter subsequent exposure of the cervical squamocolumnar junction to infectious agents and other factors. Micronutrient levels or reduced dietary intake of vitamin A, vitamin C and other micronutrients have been associated with invasive cervical cancer suggesting that some aspects of a deficient diet may contribute to these cancer outcomes [24,98].

Both positive and negative associations between human leukocyte antigen (HLA) class I and II haplotypes have been reported for both CIN and invasive cervical cancer [3,4,17,46,62,128,149,178]. The particular HLA alleles and haplo-types reported have varied between studies. Comparisons of reported HLA associations are difficult for several reasons. Differences between studies in HLA laboratory methods and HLA loci targeted have contributed but differences in ethnic composition, study design and control groups have further complicated these comparisons. In addition, particular HPV 16 variants have been associated with risk for high-grade SIL (HSIL) [175,187] and cervical cancer [74]. The most consistent data in this area has suggested that non-European (NE) HPV 16 variants have an increased risk for HSIL or invasive cancer. A study of Hildesheim and co-workers [74] found that detection of NE HPV 16 variants was associated with the presence of the HLA DRB1*1102–DQB1*0301 (two-sided P value for Fisher's exact test = 0.0005). Given historical information in other virus-host systems and the biological plausibility demonstrated by enhanced transforming abilities and differences in transcriptional regulation between HPV 16 variants, disease associations between particular HPV types or variants and specific HLA alleles or haplotypes are certain to be an area of focus for future investigations.

In the context of designing future studies in populations to address these questions, it is important to recognize certain limitations. Although HPV genotypes and their variants appear as stable entities circulating among populations worldwide, no convincing data exist to substantiate the population dynamics of particular genotypes and variants over the past 20 to 50 years. For example, we do not know if the proportion of HPV 16 variants circulating today was similar to that circulating 20 or 30 years ago and beyond. Therefore, when we use case-control study designs to evaluate HPV variant-specific risk for invasive cervical cancer, the use of current control populations infected with HPV more recently compared to women with invasive cancers who were infected with HPV 16 presumably on average 20–30 years previously may represent a highly significant bias. One might consider that the use of age-matched controls would help address this bias, however, it is not clear that older women with current HPV infections and no past history of HSIL or cervical cancer would represent the distribution of HPV variants infecting the original birth cohort of women. In addition, appropriate population representation within case-control studies needs to be achieved. Hospital or clinic-based studies suffer biases that are too significant when attempting to address these questions. With regard to HPV variant associations and HSIL including those identified recently in cohort studies, it is important to recognize that these associations are related to histological outcomes, many which are of the CIN2 category, and these observations cannot be translated into risks for invasive cervical cancer. It can be said that HPV or HLA risks associated with HSIL are morphologically defined and that infection by these HPV variants may be distinguished by these morphological changes. The risk for invasive cancer outcomes in these lesions is not absolute since many of these lesions would be expected to spontaneously regress. Although recent data may be accurate, we should take caution to further understand the areas of potential bias that can be addressed.

Pap smear screening for cervical cancer was introduced over 40 years ago. This screening appears to be the main reason for the decline in the incidence of, and mortality from invasive cervical cancer in developed countries [154]. In some case-control studies, the protection against invasive cervical cancer afforded by screening has been reported [20,47]. The evaluation of these results should consider the biases in place. Confounding is realized from the fact that women who attend clinic for Pap smear screening are at lower risk of developing the disease [148]. This appears to exist even when controlling for other risk factors. Thus, the reduction in cancer risk associated with Pap screening history in these studies would be greater than that expected in the general population and will always be related to the extent of the population receiving Pap smear screening coverage. HPV testing has been suggested as a method of screening for cervical cancer [118] but whether this will be found feasible and cost effective is unknown. Randomized clinical trials are currently on-going to evaluate HPV detection as a primary screening test for cervical cancer [81].

The molecular mechanisms by which HPVs can play a role in the carcinogenic process may be related to the physical state of the HPV DNA. This has been primarily studied in the case of HPV-related cervical disease. In benign cervical lesions, HPV genomes are often detected as extrachromosomal genomes or episomes [34,91]. In many cervical cancers, HPV DNA has been found as integrated forms although exclusively episomal forms and mixes of episomal and integrated forms have also been detected [34,91]. Viral integration has frequently been reported within the E2 open reading frame that encodes HPV transcription regulatory proteins [82,174]. It has been speculated that the loss of these E2 regulatory proteins represent a potential mechanism for deregulation of the E6 and E7 open reading frames. Both E6 and E7 proteins have functions that interact closely with the host replication and transcription regulatory machinery, namely p53 and pRB respect-ively. HPV negative cervical cancers have been reported to more frequently contain mutations in the p53 gene [33]. A direct or indirect role of integrated HPV with cellular oncogenes or tumor suppressor genes has been speculated. Elaboration of a molecular model of HPV-induced carcinogenesis is continuing to evolve and it is clear that this will represent a complex network of interactions between HPV and an array of cellular processes involved in maintenance of cellular proliferation and immune surveillance.

Anogenital HPV types have also been detected in a number of additional cancers of the anogenital tract including the vagina, vulva, penis, anal, and perianal region [78]. A percentage of oropharyngeal, tonsil, larynx and tongue cancers have also been associated with anogenital HPV types [78]. More recently squamous cell carcinomas of the skin have been associated with HPV infections [16,158]. The HPV types identified in skin cancers are cutaneous types and these have been identified in individuals who are both immunocompetent and immunocompromised. No etio-logical relationship of these HPVs and skin cancers has been established however at this time. It is interesting to note that ultraviolet light-responsive elements have been identified in the noncoding genome segment of several of these cutaneous HPV types [141].

Immunology and HPV vaccines

Papillomaviruses appear to have coevolved with their hosts and are well adapted to carry out their full life-cycle in the differentiated epithelium. The epithelium or skin represents a highly immune privileged site. The complexity of the mucosal immune system and its relationship to disease processes is currently an area of intensive investigation [64].

Little is known about the immune response to HPV during natural infection. Robust immune responses have not been observed and this is probably because HPVs have evolved elaborate mechanisms to evade host immune recognition through establishment of latent or chronic infections. Levels of viral antigens are not high and the lack of cell lysis during the HPV life-cycle limits immune recognition. Virion production is restricted to the outer most layers of the keratinized epithelium and presumably this facilitates minimal access to routine immune surveillance. Humoral immune responses appear to be directed primarily at the conformational epitopes of the L1 capsid proteins and appear to be largely type-specific and have low titer [29,89]. The use of recombinant DNA technologies to produce HPV capsids in the form of virus-like particles (VLP) has facilitated these serologic measurements that were previously not possible. HPV-specific antibodies against L1-based VLPs have been shown to be neutralizing in model systems [145] although whether these antibodies are relevant to the elimination of HPV infections is unknown. Antibodies against other HPV proteins have not been consistently observed. Antibodies to HPV 16 E6 and E7 native proteins are detected in about 50% of women with late-stage cervical cancers but they are rarely detected in women with normal or premalignant cervical cytologies [165].

In terms of specific serological reactivities to HPV capsid proteins, both IgG and IgA classes of immunoglobulin have been observed. Two studies reported by Carter and coworkers [28,29] demonstrated that many women with incident HPV infections never developed serum IgG to HPV. Seroreactivity was associated with detection of HPV 16 DNA and increased number of sex partners. The median time to sero-conversion was 8.3 months in women with incident HPV 16 infection. Among women with incident HPV infections, 59.5%, 54.1%, and 68.8% seroconverted for HPV-16, 18, or 6, respectively, during the follow-up. Transient HPV DNA was associated with a failure to seroconvert following incident HPV infection; however, some women with persistent HPV DNA never seroconverted. Antibody responses to each HPV type were heterogeneous, but several type-specific differences were found: sero-conversion for HPV-16 occurred most frequently between 6 and 12 months following DNA detection, but seroconversion for HPV-6 coincided with DNA detection. Additionally, antibody responses to HPV-16 and 18 were significantly more likely to persist during follow-up than were antibodies to HPV-6. Seroconversion can occur many months after infections and the long-term persistence of HPV type-specific antibodies has not been studied extensively but levels have been suggested to be stable over time [29,109]. In addition, IgG seropositivity is strongly correlated with lifetime number of sexual partners [182]. The overall low sensitivity of HPV VLP-based serologic tests compared to HPV DNA-based assays make serologic testing for HPV minimally useful in assessing past infection or for diagnostic purposes.

Serologic studies in men are few. Limited data suggest that serologic responses are highly gender-specific. In a study of genital wart patients [67], among asymptomatic women with HPV 6, only 22% were seropositive compared with 100% of female patients with warts. However, only 16% of male patients with warts were seropositive. When the study of male populations intensifies over the next few years, correlations of HPV genital infections with serologic responses are sure to be forthcoming.

Considerable evidence suggests that cellular immunity may be important to the eradication of HPV infections (for detailed reviews see [110,112,163]). Increased incidence of HPV-associated disease is observed in transplant patients who have been immunosuppressed or in those who have acquired human immunodeficiency virus (HIV) cell-mediated immune deficiencies. Spontaneously regressing genital warts and CIN are associated with both CD4+ and CD8+ infiltrates. Most studies have examined HPV-specific CD4+ T cell responses in peripheral blood. Little data exist on responses from tissue infiltrating T-cells or their correlates with peripheral responses. Cross sectional studies have demonstrated that women with normal cytologies and with cervical neoplasia generate peripheral blood T-helper (Th) cell responses to HPV 16 proteins, including the L1 and L2 capsid proteins as well as to the E2, E4, E5, E6, and E7 proteins [60,111,173]. Data have been somewhat conflicting and therefore no specific pattern of responses has emerged in association with progression or regression of cervical disease. Future prospective cohort studies will be needed to determine if specific immune responses can predict disease outcomes.

Evidence that T-cell responses may be disrupted or altered in HPV infections and that HPV-harboring keratinocytes may modify local immune responses has been presented ([112], for a review see [23]). HPV-transformed keratinocytes have been shown to secrete various cytokines, which may influence local immune responses. Alterations in HPV positive CIN and cervical carcinoma tissues include the identification of abnormal HLA class I transcription, transcription of HLA class I in the absence of protein expression, loss of heterozygosity in the HLA chromosomal region and dysregulation of both HLA class I and II surface expression. In addition, reduction in levels of the transporter associated with class I antigen presentation (TAP-1) have been reported. Many genes involved in the generation of immune responses are clustered on the short arm of chromosome 6 and further detailed characterizations of this chromosome in relationship to HPV natural history and associated disease outcomes are warranted.

Investigations in animal models have demonstrated that systemic immunization with the papillomavirus (PV) L1 major capsid protein in the form of self-assembling VLPs can protect against subsequent experimental challenge with the target PV type [166]. A vaccine that can prevent initial or subsequent active or persistent HPV infection could reduce the health care costs associated with abnormal Pap smear management and potentially the morbidity and mortality attributable to HPV-associated anogenital cancers. In both developing and developed countries, an effective prophylactic HPV vaccine represents a desirable candidate strategy for reducing cervical cancer incidence. Presently effective cervical cancer screening programs are costly and require a broad range of health care delivery. To be effective, these programs must conduct education programs that increase Pap smear screening in at-risk populations and state-of-the-art diagnostics must continually be implemented. In addition, programs for monitoring procedural integrity and standardization of diagnostic procedures must be maintained through regulatory bodies. These requirements would be difficult to achieve in developing countries and even if these requirements were achieved, effective follow-up and treatment programs are necessary to impact on disease outcomes. Beyond the impact that HPV vaccination could have on neoplastic disease, HPV-associated genital warts represent health problems with significant morbidity that could benefit from these prevention efforts.

Because sexual transmission is the most common mode of genital HPV transmission, prophylactic immunization will need to target young adults prior to sexual activity. Because HPV 16 represents the most common HPV type found in cytologically normal women as well as in the majority of cervical cancers it represents the primary candidate for vaccine development. Currently, clinical trials for HPV VLP-based vaccines are underway for HPV types 6, 11, 16 and 18 and these studies have included monovalent and polyvalent formulations. The safety and immunogenicity of HPV VLP-based vaccines have been recently reported for three dose series of recombinant VLPs [49,71]. Preliminary reports suggest that these vaccines may be effective although additional definitive investigations and phase III studies will be needed to truly establish this. Multiple different virological and disease endpoints will be of interest. HPV PCR-based detection for HPV types included in the vaccine preparation along with detection of a broad-range of HPVs to assess potential cross protective effects of vaccination, quantitation of HPVs since simple reduction in viral load may be adequate to impact long-term disease outcomes and effects on type-specific LSIL and HSIL outcomes will certainly be considered. If effective HPV vaccines are identified for developed countries, the question will remain as to whether these vaccines or some modification of them can be provided for safe, inexpensive and stable use in developing countries.

Summary

Over the past 15 years we have watched the unfolding of an enormous volume of information regarding HPV infections. Elaboration of the incredible genetic diversity of these viruses has been important to the development of laboratory tools for epidemiological investigations and will facilitate the future direction of studies targeting specific molecular mechanisms of disease. Through the use of these laboratory tools, HPV has been determined to be a necessary but not sufficient etiologic agent in the development of cervical cancer and other cancers of the anogenital tract. The duration of incident genital HPV infections has been partially established and this information demonstrates that most detectable HPV infections are transient. Recent observations of a second peak of cervical HPV prevalence in older women suggests the possibility that at least in some women, HPV infections may lay dormant at undetectable levels and subsequently become reactivated. The potential that older women may experience a reactivation of latent HPV infections, which may be accompanied by disease, requires further investigation. Current dogma concerning the long-term natural history of HPV infections awaits clarification by future studies. Furthermore, these future investigations remain important to appropriately characterize molecular processes within the host cell that are critical to the study of specific host immune responses to these infections.

Research-grade PCR-based HPV tests continue to be important to ongoing and future epidemiological investigations that will better define HPV incidence at various anatomic sites. Of particular interest will be the elaboration of HPV infections at extragenital sites. In this regard, the potential contribution of HPVs to skin cancer outcomes is likely to become an intensive area of study. The use of HPV assays such as HC2 in large randomized clinical trials has established HPV testing as a viable option in the management of ASCUS Pap smears. Further clinical applications of various types of HPV testing, including applications to routine screening, remain an area of intensive research. In this regard, studies that examine quantities or levels of HPV genomes and specific HPV messages are currently underway. Probably the most exciting clinically relevant development of the past decade has been the implementation of clinical trials for HPV prophylactic vaccines. To date these trials have targeted cervical HPV infections. If prophylactic HPV vaccines can prevent incident HPV infection and CIN, maintenance of long-term vaccine immunity will need to be evaluated and establishment of any potential impact on the incidence of HPV-associated invasive cancers will be determined in one or two decades following widespread implementation.

Many individuals and research groups have participated in contributing to understanding the epidemiology of HPV infections. As with the past two decades, future investigations concerning HPV infections will remain an area rich in discovery for all.

References

1. Adachi A, Yasue H, Ohashi M, Ishibashi, M. A novel type of human papilloma virus DNA from the lesion of epidermodysplasia verruciforms. Jpn J Can Res 1986; 77: 978–984.
2. af Geijersstam V, Wang Z, Lewensohn-Fuchs I, Schiller JT, Forsgren M, Dillner J. Trends in seroprevalence of human papillomavirus type 16 among pregnant women in Stockholm, Sweden, during 1969–1989. Int J Cancer 1998; 76: 341–344.
3. Apple RJ, Becker TM, Wheeler CM, Erlich HA. Comparison of human leukocyte antigen DrRDQ disease associations found with cervical dysplasia and invasive cervical-carcinoma. J Natl Cancer Inst 1995; 87: 427–436.
4. Apple RJ, Erlich HA, Klitz W, Manos MM, Becker TM, Wheeler CM. HLA DR-DQ associations with cervical carcinoma show papillomavirus type-specificity. Nat Genet 1994; 6: 157–162.
5. Aral SO, Holmes KK. Epidemiology of sexual behavior and sexually transmitted diseases. In: KK Holmes, PA Mardh, PF Sparling and PJ Wiesner (Eds), Sexually Transmitted Diseases. McGraw-Hill, New York, 1990, pp. 126–141.
6. Baay MFD, Quint WGV, Koudstaal J, et al. A comprehensive study of several general and type-specific primer pairs for detection of human papillomavirus DNAA by PCR in paraffin-embedded cervical carcinomas. J Clin Microbiol 1996; 34: 745–747.
7. Baken LA, Koutsky LA, Kuypers J, Kosorok MR, Lee SK, Kiviat NB, Holmes KK. Genital human papillomavirus infection among male and female sex partners: preva-

lence and type-specific concordance. J Infect Dis 1995; 171: 429–432.

8. Barrasso R, de-Brux J, Croissant O, Orth, G. High prevalence of papillomavirus associated penile intraepithelial neoplasia in sex partners of women with cervical intraepithelial neoplasia. N Engl J Med 1987; 2317: 916–923.

9. Bauer HM, Greer CE, Manos MM. Determination of genital human papillomavirus infection by consensus PCR amplification. In: CS Herrington and JOD McGee (Eds), Diagnostic Molecular Pathology: A Practical Approach. Oxford University Press, Oxford, 1992, pp. 131–152.

10. Bauer HM, Hildesheim A, Schiffman MH, et al. Determinants of genital human papillomavirus infection in low-risk women in Portland, Oregon. Sex Transm Dis 1993; 20: 274–278.

11. Beaudenon S, Kremsdorf D, Croissant O, Jablonska S, Wain-Hobson S, Orth G. A novel type of human papillomavirus associated with genital neoplasias. Nature 1986; 321: 246–248.

12. Beaudenon S, Kremsdorf D, Obalek S, Jablonska S, Pehau-Arnaudet G, Croissant O, Orth G. Plurality of genital human papillomaviruses: characterization of two new types with distinct biological properties. Virology 1987; 161: 374–384.

13. Beaudenon S, Praetorius F, Kremsdorf D, Lutzner M, Worsaae N, Pehau-Arnaudet G, Orth G. A new type of human papillomavirus associated with oral focal epithelial hyperplasia. Invest Dermatol 1987; 88: 130–135.

14. Becker TM, Stone KM, Alexander ER. Genital human papillomavirus infection: a growing concern. Obstet Gynecol Clin North Am 1987; 14: 389–396.

15. Becker TM, Wheeler CM, McGough NS, Parmenter CA, Jordan SW, Stidley CA, McPherson S, Dorin MH. Sexually transmitted diseases and other risk factors for cervical dysplasia among southwestern Hispanic and Non-Hispanic white women. JAMA 1994; 271: 1181–1188.

16. Berkhout RJM, Tieben LM, Smits HL, Bavinck JN, Vermeer BJ, ter Schegget J. Nested PCR approach for detection and typing of epidermodysplasia verruciformis-associated human papillomavirus types in cutaneous cancers from renal transplant recipients. J Clin Microbiol 1995; 33: 690–695.

17. Beskow AH, Josefsson AM, Gyllensten UB. HLA class II alleles associated with infection by HPV 16 in cervical cancer in situ. Int J Cancer 2001; 93: 817–822.

18. Bosch FX, Castellsague X, Munoz N, de-Sanjose S, Ghaffari AM, Gonzalez LC, Gill M, Izarzugaza I, Viladiu P, Navarro C, Vergara A, Asunce N, Guerrero E, Shah KV. Male sexual behavior and human papillomavirus DNA: key risk factors for cervical cancer in Spain. J Natl Cancer Inst 1996; 88: 1066–1067.

19. Bosch FX, Manos MM, Munoz N, Sherman M, Jansen AM, Peto J, Schiffman MH, Moreno V, Kurman R, Shah KV, International Biological Study on Cervical Cancer (IBSCC) Study Group. Prevalence of human papillomavirus in cervical cancer: a worldwide perspective. J Natl Cancer Inst 1995; 87: 796–801.

20. Bosch FX, Munoz N, de Sanjose S. Risk factors for cervical cancer in Colombia and Spain. Int J Cancer 1992; 52: 750–758.

21. Boshart M, Gissmann L, Ikenberg H, Kleinheinz A, Scheurlen W, zur Hausen H. A new type of papillomavirus DNA, its presence in genital cancer biopsies and in cell lines derived from cervical cancer. EMBO J 1984; 3: 1151–1157.

22. Brandsma J, Burk RD, Lancaster WD, Pfister H, Schiffman MH. Interlaboratory

variation as an example for varying prevalence estimates of human papillomavirus infection. Int J Cancer 1989; 43: 260–262.

23. Breitburd F, Ramoz N, Salmon J, Orth G. HLA control in the progression of human papillomavirus infections. Sem Cancer Biol 1996; 7: 359–371.

24. Brinton LA. Epidemiology of cervical cancer—an overview. In: N Munoz, FX Bosch, KV Shah and A Meheus (Eds), The epidemiology of cervical cancer and human papillomavirus. International Agency for Research on Cancer, Lyon, France, 1992, pp. 3–23.

25. Brisson J, Bairati I, Morin C, Fortier M, Bouchard C, Christen A, Bernard P, Roy M, Meisels A. Determinants of persistent detection of human papillomavirus DNA in the uterine cervix. J Infect Dis 1996; 173: 794–799.

26. Brown DR, McClowry TL, Wood K, Fife KH. Nucleotide sequence and characterization of human papillomavirus type 83, a novel genital papillomavirus. Virology 1999; 260: 165–172.

27. Burk RD, Kelly P, Feldman J, et al. Declining prevalence of cervicovaginal human papillomavirus infection with age is independent of other risk factors. Sex Transm Dis 1993; 23: 333–341.

28. Carter JJ, Koutsky LA, Hughes JP, Lee SK, Kuypers J, Kiviat N, Galloway DA. Comparison of human papillomavirus types 16, 18, and 6 capsid antibody responses following incident infection. J Infect Dis 2000; 181: 1911–1919.

29. Carter JJ, Koutsky LA, Wipf GC, Christensen ND, Lee SK, Kuypers J, Kiviat N, Galloway DA. The natural history of HPV-16 capsid antibodies among a cohort of university women. J Infect Dis 1996; 174: 927–936.

30. Chan SY, Ho L, Ong CK, Chow V, Drescher B, Durst M, ter-Meulen J, Villa L, Luande J, Mgaya HN, Bernard HU. Molecular variants of human papillomavirus type 16 from four continents suggest ancient pandemic spread of the virus and co-evolution with humankind. J Virol 1992; 66: 2057–2066.

31. Chow VTK, Leong PWF, Complete nucleotide sequence, genomic organization and phylogenetic analysis of a novel genital human papillomavirus type, HLT7474-D. J Gen Virol 1999; 80: 2923–2929.

32. Chuang TY, Perry HO, Kurland LT, Ilstrup DM. Condyloma acuminatum in Rochester, Minnesota, 1950–1978. Arch Drematol 120: 1984; 469–483.

33. Crook T, Wrede D, Vousden K. P53 point mutation in HPV negative human cervical carcinoma cell lines. Oncogene 1991; 6: 873–875.

34. Cullen AP, Reid R, Campion M, Lorincz AT, Analysis of the physical state of different human papillomavirus DNAs in intraepithelial and invasive cervical neoplasms. J Virol 1991; 65: 606–612.

35. Danos O, Katinka M, Yaniv M. Molecular cloning, refined physical map and heterogeneity of methylation sites of papilloma virus type 1a DNA. Eur J Biochem 1980; 109: 457–461.

36. de Roda Husman AM, Walboomers JMM, van den Brule AJCL, Meijer CJLM, Snijders PJF. The use of general primers GP5 and GP6 elongated at their 3' ends with adjacent highly conserved sequences improves human papillomavirus detection by PCR. J Gen Virol 1995; 76: 1057–1062.

37. de Villiers E-M, Hirsch-Behnam A, von Knebel-Doeberitz C,. Neumann Ch, zur Hausen H. Two newly identified human papillomavirus types (HPV 40 and 57) isolated from mucosal lesions. Virology 1989; 171: 248–253.

38. de-Villiers EM, Wagner, D, Schneider A, Wesch H, Miklaw HH, Wahrendorf J, Papendick U, zur Hausen H. Human papillomavirus infection in women with and without abnormal cytology. Lancet 1987; 2: 703–706.

39. de Villiers E-M. Taxonomic classification of papillomavirus. Papillomavirus (Report) 2001; 12: 57–63.

40. de Villiers E-M, Gissmann L, zur Hausen H. Molecular cloning of viral DNA from human genital warts. J Virol 1981; 40: 932–935.

41. Delius H, Hofmann B. Primer-directed sequencing of human papillomavirus types. Curr Top Microbiol Immunol 1994; 186: 13–31.

42. Delius H, Saegling B, Shamanin V, de Villiers E-M. The genomes of three of four novel HPV types, defined by differences of their L1 genes, show high conservation of the E7 gene and the URR. Virology 1998; 240: 359–365.

43. Dürst M, Gissmann L, Ikenberg H, zur Hausen H. A papillomavirus DNA from a cervical carcinoma and its prevalence in cancer biopsy samples from different geographic regions. Proc Natl Acad Sci 1983; 80: 3812–3815.

44. Egawa K, Delius H, Matsukura T, Kawashima M, de Villiers E-M. Two novel types of human papillomavirus, HPV 63 and HPV 65: comparisons of their clinical and histological features and DNA sequences to other HPV types. Virology 1993; 194: 789–799.

45. Elfgren K, Kalantari M, Moberger B, Hagmar B, Dillner J. A population-based five-year follow-up study of cervical human papillomavirus infection. Am J Obstet Gynecol 2000; 183: 561–567.

46. Ellis JRM, Keating PJ, Baird J, Hounsell EF, Renouf DV, Rowe M, Hopkins D, Duggan-Keen MF, Bartholomew JS, Young LS, Stern PL. The association of an HPV 16 oncogene variant with HLA-B7 has implications for vaccine design in cervical cancer. Nat Med 1995; 1: 464–470.

47. Eluf-Neto J, Booth M, Munoz N. Human papillomavirus and invasive cervical cancer in Brazil. Br J Cancer 1994; 69: 114–119.

48. Evander M, Edlund K, Gustafsson A, Jonsson M, Karlsson R, Rylander E, Wadell G. Human papillomavirus infection is transient in young women: a population-based cohort study. J Infect Dis 1995; 171: 1026–1030.

49. Evans TG, Bonnez W, Rose RC, Koenig S, Demeter L, Suzich JA, O'Brien D, Campbell M, White WI, Balsley J, Reichman RC. A phase I study of a recombinant viruslike particle vaccine against human papillomavirus type 11 in healthy adult volunteers. J Infect Dis 2001; 183: 1485–1493.

50. Favre M, Croissant O, Orth G. Human papillomavirus type 29 (HPV-29), an HPV type cross-hybridizing with HPV-2 and with HPV-3 related types. J Virol 1989; 63: 4906.

51. Favre M, Kremsdorf D, Obalek S, Pehau-Arnaudet G, Croissant O, Orth G. Two new human papillomavirus types (HPV54 and 55) characterized from genital tumours illustrate the plurality of genital HPVs. Int J Cancer 1990; 45: 40–46.

52. Favre M, Obalek S, Jablonska S, Orth G, Human papillomavirus (HPV) type 50, a type associated with epidermodysplasia verruciforms (EV) and only weakly related to other EV-specific HPVs. J Virol 1989; 63: 4910.

53. Favre M, Obalek S, Jablonska S, Orth G. Human papillomavirus type 49, a type isolated from flat warts of renal transplant patients. J Virol 1989; 63: 4909

54. Favre M, Obalek S, Jablonska S, Orth G. Human papillomavirus type 28 (HPV-28), and HPV-3-related type associated with skin warts. J Virol 1989; 63: 4905

55. Ferlay J, Parkin DM, Pisani P, GLOBOCAN1: cancer incidence and mortality world-wide. IARC Cancer Base No. 3. International Agency for Research on Cancer, Lyon, France. 1998.

56. Franco E, Bergeron J, Villa L, Arella M, Richardson L, Arseneau J, Stanimir G. Human papillomavirus DNA in invasive cervical carcinomas and its assocation with patient survival: a nested case-control study. Cancer Epidemiol Biomarkers Prev 1996; 5: 271–275.

57. Franco EL, Villa LL, Ruiz A, Costa MC. Transmission of cervical human papillomavirus infection by sexual activity: differences between low and high oncogenic risk types. J Infect Dis 1995; 172: 456–763.

58. Gallahan D, Müller M, Schneider A, Delius H, Kahn T, de Villiers E-M, Gissmann L. Human papillomavirus type 53. J Virol 1989; 63: 4911–4912.

59. Gassenmaier A, Lammel M, Pfister H. Molecular cloning and characterization of the DNAs of human papillomaviruses 19, 20, and 25 from a patient with epidermodysplasia verruciformis. J Virol 52: 1984; 1019–1023.

60. Gill DK, Bible JM, Biswas C, Kell B, Best JM, Puchard NA, Cason J. Proliferative T-cell responses to human papillomavirus type 16 E5 are decreased amongst women with high-grade neoplasia. J Gen Virol 1998; 79: 1971–1976.

61. Gissmann L, Diehl V, Schultz-Coulon H-J, zur Hausen H. Molecular cloning and characterization of human papilloma virus DNA derived from a laryngeal papilloma. J Virol 44: 1982; 393–400.

62. Glew SS, Duggan-Keen M, Ghosh AK, Ivinson A, Sinnot P, Davidson J, Dyer PA, Stern PL. Lack of association of HLA polymorphisms with human papillomavirus-related cervical cancer. Hum Immunol 1993; 37: 157–164.

63. Goldberg GL, Vermund SL, Schiffman MH, Ritter DB, Spitzer C, Burk RD. Comparison of cytobrush and cervicovaginal lavage sampling methods for the detection of genital human papillomavirus. Am J Obstet Gynecol 1989; 161: 1669–1672.

64. Grabbe S, Luger T. The skin as an immunological organ as well as a target for immune reaction. In: L Kater and HB de la Faille (Eds), Multi-systemic Autoimmune Diseases: An Integrated Approach. Dermatological and Internal aspects. Elsevier, Amsterdam, 1995, pp. 17–41. .

65. Gravitt P, Hakeneworth A, Stoerker J. A direct comparison of methods proposed for use in widespread screening of human papillomavirus infection. Mol Cell Probes 1991; 5: 65–72.

66. Gravitt PE, Peyton CL, Alessi TQ, Wheeler CM, Coutlee F, Hildesheim A, Schiffman MH, Scott DR, Apple RJ. Improved amplification of genital human papillomviruses. J Clin Microbiol 2000; 38: 357–361.

67. Greer CE, Wheeler CM, Ladner MB, Beutner K, Coyne MYLH, Langenberg A, Yen TS, Ralston R. Human papillomavirus (HPV) type distribution and serological response to HPV type 6 virus-like particles in patients with genital warts. J Clin Microbiol 1995; 33: 2058–2063.

68. Greer CE, Wheeler CM, Manos MM. Sample preparation and PCR amplification from paraffin-embedded tissues. PCR Methods Applications 1994; 3: S113–S122

69. Grimmel M, de Villiers E-M, Neumann Ch, Pawlita M, zur Hausen H. Characterization of a new human papillomavirus (HPV 41) from disseminated warts and detection of its DNA in some skin carcinomas. Int J Cancer 1988; 41: 5–9.

22

70. Handley JM, Maw RD, Bingham EA, Horner T, Bharucha H, Swann A, Lawther H, Dinsmore WW., Anogenital warts in children. Clin Exp Dermatol 1993; 15: 241–247.
71. Harro CD, Pang Y-YS, Roden RBS, Hildesheim A, Wang Z, Reynolds MJ, Mast TC, Robinson R, Murphy BR, Karron RA, Dillner J, Schiller JT, Lowy DR. Safety and immunogenicity trial in adult volunteers of human papillomavirus 16 L1 virus-like particle vaccine. J Natl Cancer Inst 2001; 93: 284–292.
72. Heilman CA, Law M-F, Israel MA, Howley PM. Cloning of human papilloma virus genomic DNAs and analysis of homologous polynucleotide sequences. J Virol 1980; 36: 395–407.
73. Herrero R, Hildesheim A, Bratti C, Sherman ME, Hutchinson M, Morales J, Balmaceda I, Greenberg MD, Alfaro M, Burk RD, Wacholder S, Plummer M, Schiffman M. Population-based study of human papillomavirus infection and cervical neoplasia in rural Costa Rica. J Natl Cancer Inst 2000; 92: 464–474.
74. Hildesheim A, Schiffman M, Bromley C, Wacholder S, Herrero R, Bratti MC, Sherman ME, Scarpidis U, Lin Q-Q, Terai M, Bromley RL, Buetow K, Apple RJ, Burk RD. Human papillomavirus type 16 variants and risk of cervical cancer. J Natl Cancer Inst 2001; 93: 315–318.
75. Hildesheim A, Schiffman MH, Gravitt PE, Glass AG, Greer CE, Zhang T, Scott DR, Rush BB, Lawler P, Sherman ME, Kurman RJ, Manos MM. Persistence of type-specific human papillomavirus infection among cytologically normal women. J Infect Dis 1994; 169: 235–240.
76. Ho GY, Bierman R, Beardsley L, Chang CJ, Burk RD. Natural history of cervicovaginal papillomavirus infection in young women. N Engl J Med 1998; 338: 423–428.
77. Ho L, Tay SK, Chan SY, Bernard HU. Sequence variants of human papillomavirus type 16 from couples suggest sexual transmission with low infectivity and polyclonality in genital neoplasia. J Infect Dis 1993; 168: 803–809.
78. IARC Working Group on the Evaluation of Carcinogenic Risks to Humans. IARC Monographs on the Evaluation of Carcinogenic Risks to Humans, Vol 64. Human Papillomaviruses. International Agency for Research on Cancer, Lyon, France, 1995.
79. Jacobs MV, Snijders PJF, van den Brule AJC, Helmerhorst TJ, Meijer CJ, Walboomers JM. General primer GP5+/GP6+-mediated polymerase chain reaction-enzyme immunoassay method for detection of 14 high-risk and 6 low-risk human papillomavirus genotypes in cervical scrapings. J Clin Microbiol 1997; 35: 795–805.
80. Jacobs MV, Walboomers JMM, Snijderset PFJ, et al. Distribution of 37 mucosotropic HPV types in women with cytologically normal cervical smears: the age-related patterns for high-risk and low-risk types. Int J Cancer 87: 2000; 221–227.
81. Jenkins D, Sherlaw-Jones C, Gallivan S. Assessing the role of HPV testing in cervical cancer screening. Papillomavirus Rep 1998; 9: 89–101.
82. Kalantari M, Karlsen F, Kristensen G, Holm R, Hagmar B, Johansson B. Disruption of the E1 and E2 reading frames of HPV 16 in cervical carcinoma is associated with poor prognosis. Int J Gynecol Pathol 1998; 17: 146–153.
83. Kashima HK, Shah H, Lyles A, Glackin R, Muhammad N, Turner L, van-Zandt S, Whitt S, Shah K. A comparison of risk factors in juvenile-onset and adult-onset recurrent respiratory papillomatosis. Laryngoscope 1992; 102: 9–13.
84. Kataoka A, Claesson U, Hansson BG, Eriksson M, Lindh E, Human papillomavirus infection of the male diagnosed by Southern-blot hybridization and polymerase chain

reaction: comparison between urethra samples and penile biopsy samples. J Med Virol 1991; 33: 159–164.

85. Kawashima M, Favre M, Jablonska S, Obalek S, Orth G, Characterization of a new type of human papillomavirus (HPV) related to HPV5 from a case of actinic keratosis. Virology 1986; 154: 389–394.

86. Kawashima M, Jablonska S, Favre M, Obalek S, Croissant O, Orth G. Characterization of a new type of human papillomavirus found in a lesion of Bowen's disease of the skin. J Virol 1986. 57: 688–692.

87. Kaye JN, Cason J, Pakarian FB, Jewers RJ, Kell B, Bible J, Raju KS, Best JM. Viral load as a determinant for transmission of human papillomavirus type 16 from mother to child. J Med Virol 1994; 44: 415–421.

88. Kino N, Sata T, Sugase M, Matsukura T., Molecular cloning and nucleotide sequence analysis of a novel human papillomavirus (type 82) associated with vaginal intraepithelial neoplasia. Clin Diagn Lab Immunol 2000; 7: 91–95.

89. Kirnbauer R, Hubbert NL, Wheeler CM, Becker TM, Lowy DR, Schiller JT. A virus-like particle enzyme immunosorbent assay detects serum antibodies in a majority of women infected with human papillomavirus type 16. J Natl Cancer Inst 1994; 86: 494–499.

90. Kjaer SK, van den Brule AJC, Bock JE. Determinants for genital human papillomavirus (HPV) infection in 1000 randomly chosen young Danish women with normal Pap smear: are there different risk profiles for oncogenic and nononcogenic HPV types. Cancer Epidemiol Biomarkers Prev 6: 1997; 799–805.

91. Klaes R, Woerner SM, Ridder R, Wentzensen N, Duerst M, Schneider A, Lotz B, Melsheimer P, von Knebel Doeberitz M. Detection of high-risk cervical intraepithelial neoplasia and cervical cancer by amplification of transcripts derived from integrated papillomavirus oncogenes. Cancer Res 59: 1999; 6132–3136.

92. Kleter B, van Doornter LJ, Scheggett J. Novel short-fragment PCR for highly sensitive broad-spectrum detection of anogenital human papillomviruses. Am J Pathol 1998; 153: 1731–1739.

93. Koutsky LA, Holmes KK, Critchlow CW, Stevens CE, Paavonen J, Beckmann AM, DeRouen TA, Galloway DA, Vernon D, Kiviat NB. A cohort study of the risk of cervical intraepithelial neoplasia grade 2 or 3 in relation to papillomavirus infection. N Engl J Med 1992; 327: 1272–1278.

94. Kremsdorf D, Favre M, Jablonska S, Obalek S, Rueda LA, Lutzner MA, Blanchet-Bardon C, Van Voorst Vader PC, Orth G. Molecular cloning and characterization of the geomes of nine newly recognized human papillomavirus types associated with epidermo-dysplasia verruciforms. J Virol 1984. 52; 1013–1018.

95. Kremsdorf D, Jablonska S, Favre M, Orth G. Human papillomavirus associated with epidermodysplasia verruciforms. J Virol 1983; 48: 340–351.

96. Kremsdorf D, Jablonska S, Favre M, Orth G., Biochemical characterization of two type of human papillomavirus associated with epidermodysplasia verruciforms. J Virol 1982; 43: 436–447.

97. Kurman RJ, Henson DEHAL, Noller KL, Schiffman MH, 1994. Interim guidelines for management of abnormal cervical cytology. The 1992 National Cancer Institute Workshop. JAMA 271: 1866–1869.

98. La Vecchia C, Decarli A, Fasoli M. Dietary vitamin A and the risk of intraepithelial and invasive cervical neoplasia. Gynecol Oncol 1988; 30: 187–195.

99. Lazcano-Ponce E, Herrero R, Munoz N, Cruz A, Shah KV, Alonso P, Hernandez P, Salmeron J, Hernandez M. Epidemiology of HPV infection among Mexican women with normal cervical cytology. Int J Cancer 2001; 91: 412–420.

100. Lazcano-Ponce E, Herrero R, Munoz N, Hernandez-Avila, M, Salmeron, J, Leyvaet A, et al. High prevalence of human papillomavirus infection in Mexican males: comparative study of penile-urethral swabs and urine samples. Sex Transm Dis 2001; 28: 277–280.

101. Liaw K-L, Hildesheim A, Burk RD, Gravitt P, Wacholder S, Manos MM, Scott DR, Sherman ME, Kurman RJ, Glass AG, Anderson SM, Schiffman M. A prospective study of human papillomavirus (HPV) type 16 DNA detection by polymerase chain reaction and its association with acquisition and persistence of other HPV types. J Infect Dis 2001; 183: 8–15.

102. Lo KW, Cheung TH, Chung TK, Want VW, Poon JS, Li JC, Lam P, Wong YF. 2001; Clinical and prognostic siginificance of human papillomavirus in a Chinese population of cervical cancers. Gynecol Obstet Invest 2001; 51: 202–207.

103. Londesborough P, Ho L, Terry G, Cuzick J, Wheeler C, Singer A. Human papillomavirus genotype as a predictor of persistence and development of high-grade lesions in women with minor cervical abnormalities. Int J Cancer 1996; 21: 364–368.

104. Longuet M, Beaudenon S, Orth G. Two novel genital human papillomavirus (HPV) types, HPV68 and HPV70, related to the potentially oncogenic HPV 39. J Clin Microbiol 1996; 34: 738–744.

105. Longuet, M Cassonnet, P, Orth G. A novel genital human papillomavirus (HPV), HPV type 74, found in immunosuppressed patients. J Clin Microbiol 1996; 34: 1859–1862.

106. Lorincz AT, Lancaster WD, Temple GF. Cloning and characterization of the DNA of a new human papillomavirus from a woman with dysplasia of the uterine cervix. J Virol 1986; 58: 225–229.

107. Lörincz AT, Quinn AP, Goldsborough MD, Schmidt BJ, Temple GF. Cloning and partial DNA sequencing of two new human papillomavirus types associated with condylomas and low-grade cervical neoplasia. J Virol 1989; 63: 2829–2834.

108. Lörincz AT, Quinn AP, Lancaster WD, Temple GF. A new type of papillomavirus associated with cancer of the uterine cervix. Virology 1987; 159: 187–190.

109. Luostarinen T, af Geijersstam V, Bjorge T, Eklund C, Hakama M, Hakulinen T, Jellum E, Koskela P, Paavonen J, Pukkala E, Schiller JT, Thoresen S, Youngman LD, Dillner J, Lehtinen M. No excess risk of cervical cancer carcinoma among women seropositive for both HPV 16 and HPV6/11. Int J Cancer 1999. 80: 818–822.

110. Luxton J, Shepard P. Human papillomavirus antigens and T-cell recognition. Curr Opin Infect Dis 2001; 14: 139–143.

111. Luxton JC, Rowe AJ, Cridland JC, Coletart T, Wilson P, Shepherd PS. Proliferative T cell responses to the human papillomavirus type 16 E7 protein in women with cervical dysplasia and cervical carcinoma and in healthy individuals. J Gen Virol 1996; 77: 1585–1593.

112. Majewski S, Jablonska S. Immunolgoy of HPV infection and HPV-associated tumors. Int J Dermatol 1998; 37: 81–95.

113. Makni H, Franco EL, Kaiano J, Villa LL, Labrecque S, Dudley R, Storey A, Matlashewski G. P53 polymorphism in codon 72 and risk of human papillomavirus-induced cervical cancer: effect of interlaboratory variation. Int J Cancer 2000; 87: 528–533.

114. Manos MM, Ting Y, Wright DK, Lewis AJ, Broker TR. Use of polymerase chain

reaction amplification for the detection of genital human papillomviruses. Cancer Cells 1989; 7: 209–214.

115. Matsukura T, Iwasaki T, Kawashima M. Molecular cloning of a novel human papillomavirus (type 60) from a plantar cyst with characteristic pathological changes. Virology 1992; 190: 561–564.

116. Matsukura T, Sugase M. Molecular cloning of a novel human papillomavirus (type 58) from an invasive cervical carcinoma. Virology 1990; 177: 833–836.

117. Matsukura T, Sugase M. Identification of genital human papillomaviruses in cervical biopsy specimens: segregation of specific virus types in specific clinicopthologic lesions. Int J Cancer 1995; 61: 13–22.

118. Meijer CJLM, van den Brule AJC, Snijders PFJ. Detection of human papillomavirus in cervical scrapes by the polymerase chain reaction in relation to cytology: possible implications for cervical cancer screening. In: N Munoz, FX Bosch and KV Shah (Eds), The Epidemiology of Human Papillomavirus and Cervical Cancer. International Agency for Research on Cancer, Scientific Publication No. 119. International Agency for Research on Cancer, Lyon, France, 1992, pp. 271–281.

119. Melkert PWJ, Hopman E, van den Brule AJC, Risse EKH, vanDiest PJ, Bleker OP, Helmerhorst T, Schipper MEI, Meijer CJLM, Walboomers JMM. Prevalence of HPV in cytomorphologically normal cervical smears, as determined by the polymerase chain reaction, is age dependent. Int J Cancer 1993; 53: 919–923.

120. Mitriani-Rosenbaum S, Tsvieli R, Tur-Kaspa R. Oestrogen stimulates differential transcription of human papillmavirus type 16 in SiHa cervical carcinoma cells. J Gen Virol 1989; 70: 2227–2232.

121. Moscicki AB, Palefsky J, Smith G, Siboshshki S, Schoolnik G. Variability of human papillomavirus DNA testing in a longitudinal cohort of young women. Obstet Gynecol 1993; 82: 548–585.

122. Müller M, Kelly G, Fiedler M, Gissmann L. Human papillomavirus type 48. J Virol 1989; 63: 4907–4908.

123. Munoz N, Castellsague X, Bosch FX, Tafur L, de-Sanjose S, Aristizabal N, Ghaffari AM, Shah KV. Difficulty in elucidating the male role in cervical cancer in Colombia, a high risk area for the disease. J Natl Cancer Inst 1996; 88: 1068–1075.

124. Naghashfar ZS, Rosenshein NB, Lörincz AT, Buscema J, Shah KV. Characterization of human papillomavirus type 45, a new type 18-related virus of the genital tract. J Gen Virol 1987; 68: 3073–3079.

125. Naguib SM, Lundin FE, Davis HJ. Relation of various epidemiologic factors to cervical cancer as determined by a screening program. Obstet Gynecol 1966; 28: 451–459.

126. Nuovo GJ, Crum CP, de Villiers E-M, Levine RU, Silverstein SJ. Isolation of a novel human papillomavirus (type 51) from a cervical condyloma. J Virol 1988; 62: 1452–1455.

127. Obalek S, Jablonska S, Favre M, Walczak L, Orth G. Condylomata acuminata in children: frequent association with human papillomaviruses responsible for cutaneous warts. J Am Acad Dermatol 1990; 23: 205–213.

128. Odunsi K, Terry G, Ho L, Bell J, Cuzick J, Ganesan TS. Association between HLA DQB1*03 and cervical intraepithelial neoplasia. Mol Med 1995; 1: 161–171.

129. Oriel JD. Natural history of genital warts. Br J Vener Dis 1971; 47: 1–13.

130. Orth G, Favre M, Croissant O. Characterization of a new type of human papillomavirus that causes skin warts. J Virol 1977; 24: 108–120.

131. Orth G, Jablonska S, Favre M, Croissant O, Obalek S, Jarzabeck-Chorzelska M, Jibard N. Identification of papillomavirus in Butcher's warts. J Invest Dermatol 1981; 76: 97–102.

132. Ostrow RS, Zachow KR, Shaver MK, Faras AJ. Human papillomavirus type 27: detection of a novel human papillomavirus in common warts of a renal transplant recipient. J Virol 1989; 63: 4904

133. Ostrow RS, Zachow KR, Thompson O, Faras AJ. Molecular cloning and characterization of a unique type of human papillomavirus from an immune deficient patient. J Invest Dermatol 1984; 82: 362–366.

134. Parazzini F, La Vecchia C, Negri E, Cecchetti G, Fedele L. Reproductive factors and the risk of invasive and intraepithelial cervical neoplasia. Br J Cancer 1989; 59: 805–809.

135. Parkin DM, Pisani P, Ferlay J. Estimates of the worldwide incidence of 18 major cancers in 1985. Int J Cancer 1993; 54: 594–606.

136. Parkin DM, Pisani P, Ferlay J. Estimates of the worldwide incidence of 25 major cancers in 1990. Int J Cancer 1999; 80: 827–841.

137. Peyton CL, Schiffman M, Lorincz AT, Hunt WC, Mielzynska I, Bratti C, Eaton S, Hildesheim A, Morera LA, Rodriguez AC, Herrero R, Sherman ME, Wheeler CM. Comparison of PCR- and Hybrid Capture-based human papillomavirus detection systems using multiple cervical specimen collection strategies. J Clin Microbiol 1998; 36: 3248–3254.

138. Pfister H, Hettich I, Runne U, Gissmann L, Chilf GN. Characterization of human papillomavirus type 13 from focal epithelial hyperplasia Heck lesions. J Virol 1983; 47: 363–366.

139. Pfister H, Nürnberger F, Gissmann L, zur Hausen H. Characterization of a human papillomavirus from epidermodysplasia verruciforms lesions of a patient from Upper-Volta. Int J Cancer 1981; 27: 645–650.

140. Prentice RL, Thomas DB. On the epidemiology of oral contraceptives and disease. Adv Cancer Res 1987; 49: 285–401.

141. Purdie K, Pennington J, Proby C, Khalaf S, de Villiers EM, Leigh IM, Storey A. The promoter of a novel human papillomavirus HPV 77 associated with skin cancer displays UV-responsiveness which is mediated through a consensus p53 binding sequence. EMBO J 1999; 18: 5359–5369.

142. Rho J, Roy-Burman A, Kim H, de Villiers E-M, Matsukura T, Choe J. Nucleotide sequence and phylogenetic classification of human papillomavirus type 59. Virology 1994; 203: 158–161.

143. Richardson H, Franco E, Pintos J, Bergeron J, Arella M, Tellier P. Determinants of low-risk and high-risk cervical human papillomavirus infections in Montreal university students. Sex Transm Dis 2000; 27: 79–86.

144. Roden RB, Lowy, DR, Schiller JT. Papillomavirus is resistant to desiccation. J Infect Dis 1997; 176: 1076–1079.

145. Rose RC, Reichman RC, Bonnez W. Human papillomavirus (HPV) type 11 recombinant virus-like particles induce the formation of neutralizing antibodies and detect HPV-specific antibodies in human sera. J Gen Virol 1994. 75: 2075–2079.

146. Rousseau MC, Franco EL, Villaet LL et al. A cumulative case-control study of risk factor profiles for oncogenic and nononcogenic cervical human papillomavirus infections. Cancer Epidemiol Biomarkers Prev 2000; 9: 469–476.

147. Rylander E, Ruusuvaara L, Almstomer MW, Evander M, Wadell G. The absence of vaginal human papillomaviurs 16 DNA in women who have not experienced sexual intercourse. Obstet Gynecol 1994; 83: 735–737.

148. Sasco AJ. Validity of case-control studies and randomized controlled trials of screening. Int J Epidemiol 1991; 20: 1143–1144.

149. Sastre GX, Loste MN, Salomon AV, Favre M, Mouret E, De la Rochefordiere A, Durand JC, Tartour E, Lepage V, Charron D. Decreased frequency of HLA-DRB1*13 alleles in French women with HPV-positive carcinoma of the cervix. Int J Cancer 1996; 69: 159–164.

150. Scheurlen W, Gissmann L, Gross G, zur Hausen H. Molecular cloning of two new HPV types (HPV 37 and HPV 38) from a keratoacanthoma and a malignant melanoma. Int J Cancer 1986; 37: 505–510.

151. Schiffman MH. Recent progress in defining the epidemiology of human papillomaviurs infections and cervical neoplasia. J Natl Cancer Inst 1992; 84: 394–398.

152. Schiffman MH, Bauer HM, Hoover RN, Glass AG, Cadell DM, Rush BB, Scott DR, Sherman ME, Kurman RJ, Wacholder S et al. Epidemiolgic evidence showing that human papillomavirus infection causes most cervical intraepithelial neoplasia. J Natl Cancer Inst 1993; 85: 1868–1870.

153. Schiffman MH, Bauer HM, Lorincz AT, Manos MM, Byrne JC, Glass AG, Cadell DM, Howley PM. Comparison of Southern blot hybridization and polymerase chain reaction methods for the detection of human papillomavirus DNA. J Clin Microbiol 1991; 29: 573–577.

154. Schiffman MH, Brinton LA, Devesa SS. Cervical cancer. In: D Shottenfeld and JR Fraumeni (Eds), Cancer Epidemiology and Prevention, 2nd Edn. Oxford University Press, New York, NY, 1996, pp. 1090–1116.

155. Schneider A, Kirchmayr R, de-Villiers EM, Gissmann L. Sub-clinical human papilloma-virus infection in male sexual partners of female carriers. J Urol 1988; 140: 1431–1434.

156. Schwartz SM, Daling JR, Shera KA, Madeleine MM, McKnight B, Galloway DA, Porter PL, McDougall JK. Human papillomavirus and prognosis of invasive cervical cancer: a population-based study. J Clin Oncol 2001; 19: 1906–1915.

157. Sedlacek SA, Lindheim S, Eder C, Hasty L, Woodland M, Ludomirsky A, Rando R. Mechanisms for human papillomavirus transmission at birth. Am J Obstet Gynecol 1989; 161: 55–59.

158. Shamanin V, zur Hausen H, Lavergne D, Proby CM, Leigh IM, Neumann C, Hamm H, Goos M, Haustein UF, Jung EG. Human papillomavirus infections in non-melanoma skin cancers from renal transplant recipients and nonimmunosuppressed patients. J Natl Cancer Inst 1996; 88: 801–811.

159. Shimoda K, Lorincz AT, Temple GF, Lancaster WD. Human papillomavirus type 52: a new virus associated with cervical neoplasia. J Gen Virol 1988; 69: 2925–2928.

160. Smith EM, Johnson SR, Cripe T, Perlman S, McGuinness GJD, Cripe L, Turek LP. Perinatal transmission and maternal risks of human papillomavirus infections. Cancer Detect Prev 1995; 19: 196–205.

161. Smith EM, Johnson SR, Cripe TP, Pignatari S, Turek L. Perinatal vertical transmission of human papillomavirus and subsequent development of respiratory tract papilloma-tosis. Ann Otol Rhinol Laryngol 1991; 100: 4479–4483.

162. Solomon D, Schiffman M, Tarone R, and For the ALTS Group. Comparison of three

management strategies for patients with aptypical squamous cells of undetermined significance: baseline results from a randomized trial. J Natl Cancer Inst 2001; 93: 293–299.

163. Stern PL. Immunity to human papillomavirus associated cervical neoplasia. Adv Cancer Res 1996; 69: 175–211.

164. Sugase M, Matsukura T. Distinct manifestations of human papillomaviruses in the vagina. Int J Cancer 1997; 72: 412–415.

165. Sun Y, Shah KV, Muller M, Munoz N, Bosch XF, Viscidi RP. Comparison of peptide enzyme-linked immunosorbent assay and radioimmunoprecipitation assay with *in vitro*-- translated proteins for detection of serum antibodies to human papillomavirus type 16 E6 and E7 proteins. J Clin Microbiol 1994; 32: 2216–2220.

166. Suzich JA, Ghim SJ, Palmer-Hill FJ, White WI, Tamura JK, Bell JA, Newsome JA, Jenson AB, Schlegel R. Systemic immunization with papillomavirus L1 protein completely prevents the development of viral mucosal papillomas. Proc Natl Acad Sci USA 1995; 92: 11553–11557.

167. Syrjanen K, Hakama M, Saarikoski S, Vayrynen M, Yliskoski M, Syrjanen S, Katja V, Castren O. Prevalence, incidence, and estimated life-time risk of cervical human papillomaviurs infection in a nonselecdted Finnish female population. Sex Transm Dis 1990; 17: 15–19.

168. Syrjanen S, Puranen M. Human papillomavirus infections in children: the potential role of maternal transmission. Crit Rev Oral Biol Med 2000; 11: 259–274.

169. Tang CK, Shermeta DW, Wood C. Congenital condylomata acuminata. Am J Obstet Gynecol 1978; 131: 912–913.

170. Tawheed AR, Beaudenon S, Favre M, Orth G. Characterization of human papilloma-virus type 66 from an invasive carcinoma of the uterine cervix. J Clin Microbiol 1991; 29: 2656–2660.

171. Terai M, Burk RD. Complete nucleotide sequence and analysis of a novel human papillomavirus (HPV 84) genome cloned by an overlapping PCR method. Virology 2001; 279: 109–115.

172. The ALTS Group. Human papillomavirus testing for triage of women with cytologic evidence of low-grade squamous intraepithelial lesions: baseline data from a randomized trial. J Natl Cancer Inst 2000; 92: 397–402.

173. Tsukui T, Hildesheim A, Schiffman MH, Lucci J, Contois D, Lawler P, Rush BB, Lorincz AT, Corrigan A, Burk RD, Ou W, Marshall MA, Mann D, Carrington M, Clerici M, Shearer GM, Carbone DP, Scott DR, Houghten RA, Berzofsky JA. Interleukin 2 production *in vitro* by peripheral lymphocytes in response to human papillomavirus-derived peptides: correlation with cervical pathology. Cancer Res 1996; 56: 3967–3974.

174. Vernon SD, Unger ER, Miller DL, Lee DR, Reeves WC. Association of human papillomavirus type 16 integration in the E2 gene with poor disease-free survival from cervical cancer. Int J Cancer 1997. 74: 50–56.

175. Villa LL, Sichero L, Rahal P, Caballero O, Ferenczy A, Rohan T, Franco EL. Molecular variants of human papillomavirus types 16 and 18 preferentially associated with cervical neoplasia. J Gen Virol 2000; 81: 2959–2968.

176. Volter C, He Y, Delius H, Roy-Burman A, Greenspan JS, Greenspan D, de Villiers E-M. Novel HPV types present in oral papillomatous lesions from patients with HIV infection. Int J Cancer 1996; 66: 453–456.

177. Walboomers JMM, Jacobs MV, Manos MM, Bosch FX, Lummer J A Shah, KV, Snijders PJF, Peto J, Meijer CJLM, Munoz N. Human papillomavirus is a necessary cause of invasive cervical cancer worldwide. J Pathol 1999; 189: 12–19.

178. Wank R, Thomssen C. High risk of squamous cell carcinoma of the cervix for women with HLA-DQw3. Nature (Lond) 1991; 352: 723–725.

179. Wheeler CM, Greer CE, Becker TM, Hunt WC, Anderson SM, Manos MM. Short-term fluctuations in the detection of cervical human papillomvirus DNA. Obstet Gynecol 1996; 88: 261

180. Wheeler CM, Parmenter CA, Hunt WC, Becker TM, Greer CE, Hildesheim A, Manos MM. Determinants of genital human papillomavirus infection among cytologiclaly normal women attending the University of New Mexico. Sex Transm Dis 1993; 20: 286–289.

181. Wheeler CM, Yamada T, Hildesheim A, Jenison SA. Human papillomavirus type 16 sequence variants: identification by E6 and L1 lineage-specific hybridization. J Clin Microbiol 1997; 35: 11–19.

182. Wideroff L, Schiffman MH, Hoover R, Tarone RE, Nonnenmacher B, Hubbert N, Kirnbauer R, Greer CE, Lorincz AT, Manos MM, Glass AG, Scott DR, Sherman ME, Buckland J, Lowy D, Schiller J. Epidemiologic determinants of seroreactivity to human papillomavirus (HPV) type 16 virus-like particles in cervical HPV 16 DNA-positive and negative women. J Infect Dis 1996; 174: 937–943.

183. Winkelstein Jr. W. Smoking and cervical cancer current status: a review. Am J Epidemiol 1990; 131: 945–957.

184. Woodman CBJ, Collins S, Winter H, Bailey A, Ellis JPP, Yates M, Rollason TP, Young LS, Natural history of cervical human papillomavirus infection in young women: a longitudinal cohort study. Lancet 2001; 357: 1831–1836.

185. Xi LF, Demers GW, Kiviat NB, Kuypers J, Beckmann AM, Galloway DA, Sequence variation in the noncoding region of human papillomavirus type 16 detected by single-strand conformation polymorphism analysis. J Infect Dis 1993; 168: 610–617.

186. Xi LF, Koutsky LA. Epidemiology of genital human papillomavirus infections. Bull Inst Pasteur 1997; 95: 161–178.

187. Xi LF, Koutsky LA, Galloway DA, Kuypers J, Hughes JP, Wheeler CM, Holmes KK, Kiviat NB. Genomic variations of human papillomavirus type 16 and risk for high grade cervical intraepithelial neoplasia. J Natl Cancer Inst 1997; 89: 796–802.

188. Xu S, Liu L, Lu S. Presence of human papillomavirus DNA in aminiotic fluids of pregnant women with cervical lesions. Chung-Hua Fu Chan Ko Tsa Chih 1995; 30: 457–459.

189. Yamada T., Manos MM, Peto J, Greer CE, Munoz N, Bosch FX, Wheeler CM. Human papillomavirus type 16 sequence variation in cervical cancers: a worldwide perspective. J Virol 1997; 71: 2463–2472.

190. Yamada T, Wheeler CM, Halpern AL, Stewart ACM, Hildesheim A, Jenison SA. Human papillomavirus type 16 variant lineages in United States populations characterized by nucleotide sequence analysis of the E6, L2, and L1 coding segments. J Virol 1995; 69: 7743–7753.

191. Zachow KR, Ostrow RS, Faras AJ. Nucleotide sequence and genome organization of human papillomavirus type 5. Virology 1987; 158: 251–254.

Human Papillomaviruses
D.J. McCance (editor)

Regulation of human papillomavirus gene expression in the vegetative life cycle

Loren del Mar Peña and Laimonis A. Laimins
Department of Microbiology–Immunology, Northwestern University Medical School, 303 East Chicago, Chicago, Illinois 60611, USA

Introduction

Human papillomaviruses (HPVs) infect epithelial or mucosal tissue at specific anatomical locations. Over 100 HPVs have been identified to date, of which approximately 30 types infect the genital tract [53]. Genital types are divided into two categories based on their potential for malignant transformation. Infection by low-risk viruses, such as HPV-6, and 11, can lead to genital warts, while high-risk types (HPV-16, 18, 31, 33, and 45, among others) can induce oncogenesis in infected tissue of the anogenital tract [68].

HPVs are members of a small group of viruses that link their life-cycle to the differentiation status of the host cell. Although the mode of entry has not been definitively characterized, it is likely to occur through microwounds of the epithelium and to utilize a heparin-like molecule on basal cells for the initial binding event [60]. Following entry, viral DNA is established as episomes in basal cells at approximately 50–100 copies per cell. Early viral gene expression occurs at low levels at this stage, and viral DNA replication is synchronized to the host chromosomes. As infected cells migrate through the suprabasal strata, they differentiate and the viral DNA continues to replicate. Late gene expression occurs upon terminal differentiation, and progeny virions are produced in suprabasal cells [10,78].

The use of organotypic raft culture and other methods for differentiation has facilitated analysis of transcription and DNA synthesis throughout the HPV life-cycle [73], while growth of submerged monolayer cultures is a model for growth in the basal layers of the epithelium. To induce differentiation in organotypic culture, cells are first placed on a collagen plug containing fibroblast feeders, and the plug is transferred onto a wire grid that is maintained on an air–liquid interface. In this system, HPV-infected cells derived from biopsy specimens exhibit histological changes similar to those observed *in vivo* in lesions and can be used to induce virion production [72]. Raft cultures have also been used to successfully reproduce the complete HPV-16, 18, and 31 life-cycle from transfected DNA templates [39–41,74]. An alternative method involves suspension in semi-solid media, which has recently been reported to induce rapid differentiation. Markers of differentiation, such as transglutaminase and involucrin, are observed within 24 hours of incubation in methylcellulose. This system can induce genome amplification and activation of the late promoter, which are dependent on epithelial differentiation [91].

HPVs consist of circular double-stranded DNA of approximately 8 kb in length. Transcription, which occurs from a single strand, is largely controlled by elements within a non-coding region that consists of binding sites for a number of cellular and viral transcription factors [53]. This 1 kb region is alternatively referred to as the

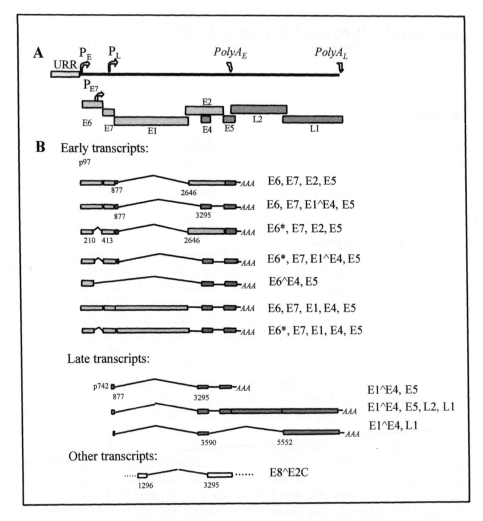

Fig. 1. (A) Linear map of the genital HPV genome. P_E designates the early promoter; P_L designates the late promoter; P_{E7} indicates the E7 promoter in low-risk viruses. PolyA$_E$ indicates the early polyadenylation site between the E5 and L1 ORF; PolyA$_L$ indicates the late polyadenylation site downstream of the L2 ORF. The upstream regulatory region is labeled URR. Late genes are indicated by dark shading. (B) Major HPV-31 transcripts that have been identified to date. Splice sites are indicated at the junctions, and the coding potential of each transcript is located to the right of each message.

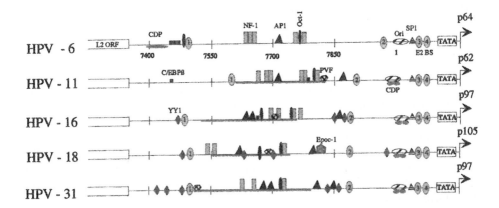

Fig. 2. Viral and cellular transcription factor binding sites in the URR of five HPVs. The core enhancer for each type of virus is shaded underneath the corresponding sequence. The location of the early promoter is indicated to the right, by arrows. E2 binding sites are indicated by numbered ovals.

non-coding region, the long control region, and the upstream regulatory region (URR). We will refer to it as the URR throughout the chapter.

Papillomaviruses encode at least eight viral gene products that follow a temporal pattern of expression (Fig. 1). Early gene products are expressed at the initial stages of infection in the basal stratum and include E1, E2, E6, and E7 [63,94]. These proteins are involved in viral replication, copy number maintenance, and disruption of the cell cycle. E1 is an ATP-dependent helicase that binds to the viral origin of replication, an A/T-rich sequence within the URR [3,25,36,55,59,69]. The protein has low binding affinity for the origin; however, a complex of E1 and E2 functions to load E1 to the origin [12,19–21,26,28,30,40,46]. In addition to its function in replication, E2 is a transcription factor that binds as a homodimer to the palindromic sequence $ACCN_6GGT$ [5,15,26,37,38,67,92]. The URR of genital papillomaviruses contains four binding sites for E2: binding site 1 is located at the 5′ end of the URR, two binding sites flank the E1 binding site at the origin of replication, and binding site 4 is located directly upstream of the early promoter (Fig. 2). E2 has differential binding affinity to each site, and it has been suggested that such differential affinity is the basis for regulation of viral gene expression [102]. The biology of E6 and E7 is discussed in other chapters. Late gene expression occurs upon epithelial differentiation and is accompanied by amplification of the viral genome.

General organization of the URR and early promoter

All HPVs contain an early promoter that is active at the initial stages of infection and is dependent on enhancer sequences present at the URR (Fig. 2). In transient transfection assays, this enhancer exhibits a preference for activation in epithelial cells. The early promoter is referred to as p97 in HPV-16 and 31 [56,88], p105 in

HPV-18 [93], and p62 in HPV-11 [96]. The early promoter in all genital types contains a TATA box approximately 30 nucleotides upstream of the transcript initiation site. Two binding sites (BS3 and BS4) for the viral protein E2 are located directly upstream of the TATA box, and a binding site for Sp1 is located 5' of E2 BS 3. These three elements are important for early promoter activation and will be discussed in detail in subsequent sections. A second early promoter exists within the E6 open reading frame of low-risk viruses HPV-6 and 11, and its TATA box maps to nucleotide 243 in HPV-6b and nucleotide 229 in HPV-11 [96]. Transcripts initiating from this low-risk promoter encode E7 as the first ORF. This contrasts with early transcripts from the high-risk viruses, where E6 or a spliced variant form of E6, called E6*, precedes E7. In addition, p229 and p243 appear to be regulated by the URR independently from the upstream promoter and may provide a mechanism for high-level expression of E7.

The URR of all HPVs consists of the origin of viral replication and a number of binding sites for viral and cellular transcription factors. Many of the binding sites correspond to proteins that are ubiquitously expressed, and it has therefore been suggested that the epithelial tropism of HPVs may be due to combinatorial binding of ubiquitous factors at the URR. Alternatively, factors expressed exclusively in epithelial tissue may be responsible. The HPV URR can be divided into three regions: the 5' distal region, a central enhancer, and a promoter-proximal region. The 5' region of the URR has not been assigned a significant function, although it contains the late transcript polyadenylation site and several motifs that appear to be important for RNA stability. In addition, a recent study suggests that the 5' URR may have a negative effect on viral DNA replication [54]. The central enhancer consists of a core element that is indispensable for early gene expression, and auxiliary binding sites that synergize activation from the core enhancer. Finally, the promoter-proximal region contains the origin of viral replication, two binding sites for E2, a Sp1 site, and the TATA box for the early promoter.

Transcription factors that control early gene expression

The URR of HPVs shows remarkable functional conservation, as many transcription factor binding sites are present at similar locations (Fig. 2). Promoter activity is likely to be mediated by complex protein–protein interactions, as well as by competition for binding. Interestingly, the importance of each factor appears to vary among different HPV types. This variation may be due to enhancer function among different cell lines, or to the methods used to study expression. The importance of Sp1 as an activator of the early promoter, however, is consistent among all types studied. Sp1 binding sites are found directly upstream of E2 binding site number 3 in the promoter proximal region of all genital HPV types. Binding of Sp1 to this site is believed to recruit the preinitiation complex to assemble at the early promoter. Mutation of this site has been shown to strongly decrease early promoter activity across cell types in transient assays. Additionally, Apt and colleagues observed that different members of the Sp family of transcription factors bind to URR sequences in HPV-16 and exert distinct effects [7]. Increased levels of Sp1, a transcriptional activator, were associated with tissues where the HPV 16 enhancer showed strong activity, while decreased URR activity was observed in cells where Sp3, a repressor, was highly expressed. It has been suggested that Sp3 competes with Sp1 for binding to the conserved site upstream of the early promoter. Changes in the ratio of Sp1 to Sp3 in various epithelial tissues therefore may act to moderate URR activity.

Another important activator of the early promoter is the activating protein-1 (AP-1). AP-1 complexes function in cell proliferation, differentiation, and survival, and consist of homo- or heterodimers of the Jun, Fos, or other families. The URR of genital HPVs contains several binding sites for AP-1, although their contribution to promoter activity varies. Mutation of the distal AP-1 binding site at nucleotide 7645 in HPV-31 resulted in a 90% decrease of reporter activity, while mutation of the proximal binding site at nt 7683 had a less severe effect [66]. In addition, AP-1 binding to the proximal binding site also appeared to interact with Oct-1 at an adjacent site. While mutation of both AP-1 and Oct-1 sites resulted in a modest increase in reporter activity, a single mutation at the Oct-1 site dramatically increased expression [66]. This data suggests a repressive role for Oct-1. AP-1 also appears to interact with other factors at both the 5' and 3' flank of the HPV-18 URR [17]. In addition, the composition of AP1 dimers changes as epithelial cells differentiate, which may modulate HPV gene expression in suprabasal cells.

All genital HPVs contain at least four non-palindromic binding sites for nuclear factor-1 (NF-1) dispersed throughout the URR, and some reports suggest that NF-1 may play an important role in early promoter activity [45]. The NF-1 family consists of many members with a common N-terminal dimerization domain and a variable transactivation domain that results from differential splicing. The many isoforms of NF-1 exhibit a tissue-type specificity that may be in part due to dimerization to different family members. Mutation of various NF-1 sites in HPV-16 and 11 led to a marked decrease in enhancer activity [22,23,31], though another group has not seen these effects in HPV-18 [17]. Interestingly, all genital HPVs contain a conserved NF-1 site within two nucleotides of a binding site for Oct-1, and Oct-1 binding at this site appears to synergize with NF-1 activation. Reporter assays indicate that increasing the spacing between the NF-1 and Oct-1 sites in HPV-16 abolished the synergistic effect in a manner similar to the effect of mutations to either the Oct-1 or NF-1 site [79]. This study suggests a role for the Oct-1/NF-1 complex in URR enhancer activity of high-risk viruses. Interestingly, no effect was observed upon mutation of the two NF-1 sites flanking the Oct-1 binding site in HPV-11 [113]. This may be due to differences in promoter regulation among low and high-risk viruses.

The octamer binding Oct-1 transcription factor seems to exert a repressive role on the early promoter in the absence of interaction with other factors. Although one report suggests that a mutation in HPV-18 had no effect on enhancer activity [17], at least three other studies in HPV-16, 18, and 31 report increased promoter activity upon ablation of the Oct-1 binding site. Mutation of an Oct-1 site around nt 7665 in HPV-16, independently of other sites, increased enhancer activity [95]. In the case of HPV-31, mutation of an Oct-1 site adjacent to an AP-1 site also led to an increase in reporter activity [66]. Furthermore, overexpression of Oct-1 in co-transfectants with HPV-16 and 18 enhancer constructs led to decreased promoter activity [51,76]. A repressive effect was not observed in HPV-11 constructs, again suggesting differences in early promoter regulation between low and high-risk viruses [114].

All genital HPVs contain at least one binding site for the CAAT enhancer binding protein β (C/EPBβ). The C/EBP family of transcription factors consists of several members that are generated by translation from different AUG codons. C/EBP proteins bind as homodimers or heterodimers with other family members or with other factors. The C/EBP proteins have been suggested to have an important role in modulation of terminal differentiation, with a well-characterized function in hepato- and adipogenesis. One study found that treatment of CaSki cells with IL-6, an inducer of C/EBPβ expression, repressed HPV-16 early gene expression [65]. Similarly, overexpression of C/EBPβ inhibited HPV-11 genome maintenance [111]. Finally, the effects of depleting C/EBPβ in HFKs were studied by use of oligonucleotides with closed ends shaped like dumbbells. Co-transfection of C/EBPβ dumbbells and HPV-11 genomes increased both copy number and E6/E7 transcripts with respect of HPV-11 transfectants alone [111].

Another important regulator of HPV expression is ying yang-1 (YY1). YY1 may activate or inhibit transcriptional activity depending on the context of the promoter. A YY1 binding site is present downstream of the C/EBPβ site in HPV-18. Previous observations suggested that YY1 binding to this downstream site was necessary for transcriptional activity of the early promoter. However, YY1 became a transcriptional repressor in the absence of upstream sequences, suggesting a complex interaction between YY1 and other transcription factors binding to the URR [8,9]. YY1 has additionally been shown to repress E6 and E7 expression in HPV-16. Although the exact mechanism of YY1 repression has not been elucidated, it may act by inhibiting AP-1 transactivation [81]. An AP-1 binding site overlaps five YY1 sites in the enhancer of HPV-16, and mutation of the YY1 sites led to a four to six-fold increase in p97 promoter activation. YY1 was found to interact with CREB binding protein (CBP), a co-activator for AP-1, and may target CBP to mediate its repressive effects. Similar arrangements of YY1 and AP-1 sites at equivalent locations are present in HPV-18, 31, and 35, suggesting a conserved mechanism for modulation of AP-1 function. YY1 also plays a negative role in the regulation of E1 promoter in HPV-6 (p680) [1]. p680 is active only upon differentiation and is homologous to the late promoter in high-risk viruses. Mutation of the YY1 site in the E1 promoter led to an increase in reporter activity in undifferentiated keratinocytes, indicating that YY1 repressed p680 activity in monolayer cells. YY1 was recently shown to interact with proteins that have histone deacetylase activity, which provides a direct link to its function as a repressor.

An additional interaction between YY1 and E2 has been reported to occur proximally to the origin of replication in HPV-18 [70]. Addition of YY1 repressed E1/E2-mediated replication of HPV-18 in a cell-free system. DNA binding by YY1 was not necessary for this negative effect, since neither competitor YY1 binding sites nor antibodies to the DNA binding domain of YY1 were able to alleviate repression.

While YY1 binds to co-repressor complexes containing histone deacetylases, de-acetylase activity seems minimally important in a cell-free system where DNA is generally free of histones. Therefore, interference with E2 function may likely be the mechanism of action.

Additional factors have been reported to be essential for HPV gene expression. The differentiation-dependent CCAAT Displacement Protein (CDP) has been shown to negatively regulate the E6 promoter in HPV-6 through a binding motif at the 5′ end of the URR of HPV-6 and a HPV-6 variant called HPV-6$_{W50}$ [2,84]. The variant HPV-6b, in contrast, contains a 94 nt deletion that spans the CDP silencer. Introduction of the silencer element from HPV-6$_{W50}$ into the HPV-6b URR was sufficient to induce repression [84]. CDP also appears to bind to sequences spanning the origin of replication in HPV-16, 31, 33, 18, 45, and 11 and to repress replication and transcription from the E6 promoter [80]. While CDP functions as a transcriptional repressor in undifferentiated cells, it is inactive in cells that have undergone differentiation. Thus, CDP has the potential to link viral replication and transcription to the life-cycle of the infected cell. The repressor function of CDP may stem from recruitment of histone deacetylase 1 and chromatin remodeling factors such that the template DNA is less accessible to the transcriptional machinery [77]. An additional factor, designated the papillomavirus enhancer factor (PVF), binds to the URR of HPV-16 [23]. Mutation of the PVF binding site led to a 45% decrease in enhancer activity. The binding site for PVF consists of an octamer-like sequence and is highly conserved among high-risk HPVs but is absent in low-risk viruses. The identity of this factor has not been conclusively determined, but it is alternatively referred to as TEF-1 and 2 in HPV-16 [7,58,79], or F5 in HPV-31 [66]. Finally, another motif that appears to be a conserved regulator of HPV gene expression is a glucocorticoid responsive element in the URR of HPV-18, 16, and 11 [17,18,75,113]. The element has been reported to be responsive to dexamethasone and to function as an enhancer. Whether it actually plays an important role in the HPV life-cycle has yet to be determined.

A recent report by Bouallaga and colleagues has provided a three-dimensional perspective on transcription factor assembly at the URR by describing the assembly of a JunB/Fra-2/HMG-I(Y) enhanceosome complex on the HPV-18 enhancer in HeLa cells [14]. Enhanceosomes consist of nucleoprotein complexes that exhibit stereospecific assembly in order to create an activating surface for co-activators and the transcriptional machinery [71]. The specific three-dimensional arrangement of ubiquitous factors is the basis of activation of a specific promoter, thereby allowing precise control over expression of different genes. The core enhancer of HPV-18 consists of a 109 bp fragment of the URR that includes an AP-1 site. The study found that internal deletions or insertions to the enhancer affected transcriptional activation, which was reflected by activity of a downstream reporter. Half-turn internal deletions to the core enhancer abolished reporter activity, while full-turn deletions, which preserved the stereo-alignment of the binding sites, did not. A rapidly migrating AP-1 complex, composed of JunB and Fra-2, bound at the enhancer at low concentrations of nuclear extract. At higher protein concentrations, a slowly migrating complex formed at the expense of the AP-1 complex and showed coopera-tivity over a narrow range of concentrations. The slow AP-1 complex contained HMG-I(Y), an architectural protein that aids in DNA bending. The core enhancer of HPV-18 exhibited similar features to the well studied interferon β and T-cell receptor-alpha enhanceosomes: stereospecific activity, a higher-order DNA-protein structure, and presence of an architectural protein that cooperatively binds to AP-1. Enhanceosome assembly on HPV-18 therefore provides an integrated view of the nucleoprotein complex that mediates activation of the early HPV promoter.

Transcription factors that may impart tissue specificity

An important question in HPV biology is what determines the epithelial tropism of the virus. Numerous studies have demonstrated that the HPV enhancer activates expression preferentially in epithelial cells. A potential explanation for this is that epithelial-specific factors activate enhancer function. One potential candidate is a POU domain protein designated the epidermal octamer-binding factor 1 (Epoc-1), which is present exclusively in differentiated epithelium [112]. Epoc-1 binds to the promoter-proximal region of the URR of HPV-18 and activates the early promoter. In addition, exogenous expression of Epoc-1 in non-epithelial cells where the HPV-18 URR is inactive can restore early promoter activation. Epoc-1 binding sites partially overlap AP-1 sites, and thus Epoc-1 binding has also been suggested to regulate AP1 binding. AP-1 is an additional example of a transcription factor with tissue-specific expression, since the tissue distribution of jun, fos, and other members of the family varies. In undifferentiated cells, the major AP-1 complex consists of JunB and Fra-2. On the other hand, c-jun, JunB, c-Fos, and Fra-2 are highly expressed at the stratum granulosum and target a number of genes involved in epithelial differentiation [4].

Additional candidates for imparting tissue specificity in HPV are transcriptional enhancer factor-1 (TEF-1) and CAAT-binding transcription factor-1 (CTF-1). Four TEF-1 binding sites have been identified in the URR of HPV-16 [58]; in addition, TEF-1 requires a co-activator that is present in extremely limiting amounts. Avail-ability of the co-activator may therefore modulate TEF-1 activity in different cell types. Another epithelial factor that has been correlated with URR activity is CTF-1, a member of the NF-1 family of transcription factors. Specific CTF-1 band shifts in epithelial cells have been associated with transcriptional activity of the URR in HPV-16 [6]. Furthermore, overexpression of CTF-1 conferred URR-mediated transcriptional activation to non-epithelial cells. Identification of other cell-type specific factors responsible for HPV gene expression in epithelial cells will require additional study.

E2 and viral gene expression

Papillomaviruses encode a small number of factors that can directly regulate viral gene expression. The most extensively characterized of these is the E2 protein. The

E2 proteins are highly conserved among all papillomaviruses and recognize a consensus palindromic site with the sequence $ACCN_6GGT$ [11]. The carboxyl-terminus of E2 contains the dimerization and the DNA-binding domain. Crystallization studies have demonstrated that the C-termini of E2 dimers form a beta-barrel structure [16,49], that is similar to one in the unrelated protein EBNA-1 from Epstein–Barr virus. The amino-terminus of E2 encodes a transactivation domain as well as a domain for interaction with E1. The N-terminus has recently been crystallized and shown to consist of alpha-helices that can dimerize. Such dimerization may facilitate DNA looping between E2 molecules bound to different DNA sequences [5,15,48]. A central hinge region connects the amino and carboxyl termini. The hinge region exhibits a high degree of sequence variability among papillomaviruses, and its structure remains to be defined.

The BPV-1 E2 transactivator can bind to seventeen sites scattered throughout the genome [34] and functions as the major activator of BPV-1 early gene expression. In contrast, only four E2 binding sites are present in the URR of genital papillomaviruses (Figure 2). Reporter assays involving the URR of HPV-16 and 18 revealed that E2 activated expression at low concentration, yet repressed it when expressed at high levels [102,107]. A possible mechanism for this repression is through displacement of TBP and Sp1 when E2 binds to E2 binding sites 3 and 4 [32,107]. A recent study suggests that E2 from HPV-11 may also repress the early promoter by directly targeting several components of the transcriptional machinery [52]. In contrast, the E7 promoter in HPV-6a is activated by E2, particularly through binding site 1 [86]. This regulation may contribute to high levels of E7 expression in the low-risk viruses. Interestingly, the HPV E2 protein, like the BPV-1 protein, can activate expression of heterologous promoters consisting of multimerized E2 binding sites [103]. It therefore seems that the particular arrangement of E2 binding sites with respect to nearby promoters leads to different modes of regulation. It is believed that E2 repression is part of a mechanism for control of viral copy number in undifferentiated cells. At low levels of protein, E2 activates E1 expression from the early promoter, while at high levels, E2 represses expression. This regulation contributes to the maintenance of constant numbers of viral template in undifferentiated cells. Upon differentiation, the late promoter directs E1 and E2 expression. Since the late promoter is not subject to regulation by E2, high-level expression of E1 and amplification of viral DNA result at the suprabasal layers of the epithelium.

Recent data suggests that a short form of E2, called E8^E2C, can mediate transcriptional repression of the early promoter of HPV-31 [104]. Alternative splicing in the E1 and E2 ORFs generates the short E8 ORF fused to the C-terminus of E2 (Fig. 1). The E8^E2C protein can interfere with the replicative function of E2, perhaps by formation of heterodimers with the full-length protein. E8^E2C has additionally been shown to play a role in early transcript repression in a novel manner. While full-length E2 repressed by binding to the promoter-proximal binding site 4, E8^E2C can act at a distance through the distal E2 binding site 1 [105]. E8^E2C-mediated repression requires E2 binding sites and also occurs with heterologous promoters with upstream E2 binding sites. Similar E8^E2C proteins have been described in HPV-11, 16, and 33 [33,89,97]. E8⌃E2C may therefore be part of a negative feedback loop that downregulates early gene expression and DNA synthesis.

The late promoter

Late gene expression requires epithelial differentiation, which serves to link the viral and cellular life-cycle. Late genes include E1⌃E4, a cytoplasmic protein thought to aid in viral egress from infected cells, and the two capsid proteins L1 and L2 [57]. The late promoter consists of a number of initiation sites that map to sequences within the E7 ORF in HPV-6, 16, and 31 [27,47,50,57,61,78,110]. There are several models for late promoter activation. The simplest model suggests that activation of the late promoter requires expression of differentiation-specific transcription factors. A second model proposes that late gene expression requires genome amplification for a gene dosage effect. Replication could also act to increase the accessibility of the DNA to the transcriptional machinery, or to titrate away an inhibitory factor. The requirement for genome amplification in activation of the late promoter is consistent with two previous observations. First, Frattini et al. showed that HPV-31 episomal templates are required for late gene expression [42]. Second, CIN612 cells that spontaneously amplify viral DNA in monolayer cultures express late genes in the absence of differentiation [85].

No consensus TATA boxes are found within the late promoter region, although potential initiator elements are present in HPV-31. Initiator elements generally overlap transcript start sites and function to recruit the transcriptional machinery. However, many of the start sites for late transcripts in HPV-31 lack initiator elements. Transcription from TATA-less and initiator-less promoters may be mediated by transcription factors that direct the transcriptional machinery to the correct site. A number of binding sites for cellular transcription factors are present in the vicinity of the late promoter. These include sites for Oct-1, SOX 5, and SRY, which bind to late promoter sequences *in vitro* (L. Peña, unpublished results); however, it is not clear if the sites function as transcriptional *cis* elements that direct late gene expression. Initiation from promoters lacking a TATA box and an initiator element occurs promiscuously over hundreds of nucleotides throughout the promoter [90]. This has been observed for late promoter transcripts of HPV-31, which occur throughout 200 nucleotides within the E7 ORF. Over 30 initiation sites have been mapped in HPV-31, including major start sites at nucleotides 626, 642, 680, 737, 742, and 767 [27,47,82,83]. Only a limited number of factors have been functionally implicated to regulate late protein expression. Among these is CDP/Cut, which has been shown to control the differentiation-specific E1 promoter in HPV-6 [2]. CDP, a transcriptional repressor in undifferentiated cells, is inactive in cells that have undergone differentiation. CDP may thus link late gene expression to differentiation status of the infected cell.

Eukaryotic DNA is densely packaged into chromatin, which consists of a nucleosomal complex of DNA and histones that is assembled into higher-order structures.

Chromatin must be modified to make DNA accessible to the transcriptional machinery [13]. Papillomavirus DNA assembles into nucleosomes; furthermore, Stünkel and Bernard have shown that chromatin rearrangement plays a role in HPV early gene expression [106]. *In vivo* studies of CaSki cells, which carry approximately 500 copies of integrated HPV-16, indicate that nucleosome positioning occurs over the viral enhancer, the origin of viral replication, and the early promoter. A nucleosome was found to be located directly over the early promoter of HPV-16 and was shown to inhibit early gene expression. A similar arrangement was observed *in vitro* over the early promoter of HPV-18. These observations suggest that chromatin architecture on a viral template may be directed by the DNA sequences and may regulate viral gene expression. The DNAse I hypersensitivity assay is helpful in assessing the state of chromatin over a DNA template, since areas where chromatin is relaxed are more accessible to digestion by DNAse I. By means of this assay, it was observed that a shift in the state of chromatin occurred during the life-cycle of HPV-31 [27]. The DNA around the late promoter became more accessible to DNAseI upon differentiation. This indicates that the chromatin structure over the late promoter becomes transcriptionally accessible at this stage of the life-cycle. Two cellular processes which can mediate chromatin rearrangements, namely histone acetylation by histone acetyltransferases (HATs), and ATP-dependent chromatin remodeling could be involved in altering chromatin structure. While HATs were found to play a role in regulation of early gene expression, these studies indicated that a mechanism other than histone acetylation was responsible for chromatin rearrangement around the late promoter [27]. Interestingly, E1 appears to interact with human SNF5, a member of the SWI/SNF family of proteins [69]. The E1/SNF5 complex may participate in modulation of viral DNA synthesis, but whether this complex functions in late promoter activation remains unclear.

Post-transcriptional regulation

Although most work on papillomavirus gene expression has focused on transcriptional regulation, a growing body of evidence suggests that post-transcriptional events may play an important role in viral gene expression. Eukaryotic pre-mRNAs are processed at the 3′ end by endonucleolytic cleavage at a GU-rich element and subsequent addition of a poly (A) tail at the cleavage site (Fig. 3A) [24]. The distance between the AAUAAA motif and the GU-rich element determines the site of cleavage and the efficiency of polyadenylation. HPV genomes contain a functionally conserved polyadenylation signal for both early and late transcripts (Fig. 1A) [109]. The early polyadenylation site is located between the E5 and L2 ORFs, and early transcript polyadenylation occurs in a heterogeneous manner over a stretch of 100 nucleotides encompassing the polyadenylation signal. Early polyadenylation motifs in HPV-31 were observed to be approximately 20-fold less efficient than the well characterized elements in SV40 [108]. Late transcripts, on the other hand, are processed more precisely, with 3′ ends extending over only 5 nts downstream of the consensus hexanucleotide motif. Substituting the early signal for late sequences

A   mRNA Processing

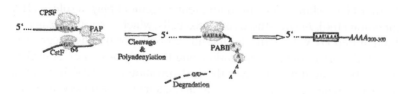

B   HPV - 31 polyadenylation

C   Ig heavy chain switch

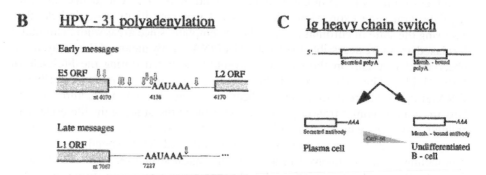

Fig. 3. (A) Schematic representation of mRNA processing. CPSF recognizes the AAUAAA motif and CstF recognizes the G/U motif, which subunit 64 binds. The proteins recruit other members of the polyadenylation machinery, which cleaves the precursor RNAs. The polyA polymerase (PAP) then adds a poly(A) tail. The poly (A) binding protein II (PABII) restricts the tail to a 200 to 300 nt length. (B) Early transcripts in HPV-31 are polyadenylated over a 100 nt region, while late transcripts are processed over a span of 5 nts. (C) The immunoglobulin (Ig) heavy chain switch is associated with increased CstF-64 levels in undifferentiated B-cells. Immature B-cells express membrane-bound Ig, while plasma cells express a secreted form. At high CstF-64 concentrations, RNA processing occurs at a weak polyadenylation site that results in secreted Ig in activated B-cells.

resulted in early transcript processing over a smaller range of viral sequence, indicating that the RNA processing elements in the viral genome are responsible for the differences in early and late mRNA processing.

RNA processing occurs in several steps [24]. The hexanucleotide motif is recognized by the cleavage and poly A specificity factor (CPSF), and the GU-rich element is recognized by the cleavage stimulation factor (CstF). CstF, in turn, increases the CPSF binding affinity to the AAUAAA motif. The CstF-64 subunit in the CstF complex binds RNA through a RNA binding domain, and CstF-64 abundance influences mRNA processing. The switch from membrane-bound to secreted forms of immunoglobulin is a well studied example of how protein abundance influences polyadenylation (Fig. 3C) [35]. Resting B-cells have low levels of CstF-64 and primarily express the membrane-bound form of IgM, which is polyadenylated at a strong site downstream of a weak one. Upon cell activation, CstF-64 levels increase, allowing the cleavage factor to recognize the weak polyadenylation site in the heavy chain gene. This results in an Ig that lacks a transmembrane domain and is secreted. Terhune et al. found that levels of CstF-64 decreased upon epithelial differentiation and this change may influence the choice of polyadenylation sites in HPVs [109].

Several sequences within the L2 and L1 ORFs and the 3′ untranslated region (3′UTR) have been shown to influence viral mRNA stability. The first 800 nt in the L2 ORF in HPV-31 contain a sequence that inhibited usage of the late polyadenylation site [108], and a similar regulatory element in HPV-16 decreased the half-life of viral transcripts [99]. Experiments using cycloheximide revealed that protein synthesis was necessary for inhibition in HPV-16, suggesting that a cellular factor was involved in mRNA degradation. A second instability element is present at the 3′ end of the L1 ORF and extends into the 3′UTR of HPV-16 [29, 62, 64]. Similar to its counterpart in L2, this L1 element was found to be capable of decreasing the half-life of mRNA transcripts. HPV-1 and BPV-1 also contain an inhibitory element in the late gene 3′UTR, downstream of the L1 stop codon [44, 98]. Two motifs are important for this function in HPV-1: a 5′ sequence that contains two AUUUA motifs, and a 3′ sequence with three UUUUU motifs [98]. The inhibitory region also includes four splice donor-like sequences and is recognized by several proteins, including several that recognize intronic sequences. The late 3′ UTR in BPV-1 also contains an instability element that reduces the levels of cytoplasmic L1 mRNA without affecting the half-life of the transcripts [43,44]. Interestingly, the inhibitory region contains a sequence that is homologous to a splice donor site. Mutations to this pseudo-splice site were found to abolish inhibitory activity and stimulated polyadenylation at the late site, perhaps by impairing the definition of an exon [44]. These examples offer compelling evidence that splicing and polyadenylation influence gene expression throughout the PV life-cycle.

In contrast to the majority of eukaryotic messages, HPV proteins are expressed from polycistronic messages. Figure 1B depicts many of the HPV-31 mRNAs that have been characterized to date. In polycistronic messages, translation of downstream cistrons occurs through several mechanisms, including ribosomal shunting and leaky scanning. In ribosomal shunting, ribosomes bind to the 5′ end of the message, travel a short distance, and then translocate to a downstream site without reading the intervening sequence. Shunting has been proposed to function in E1 translation from a polyscistronic message that encodes E6 and E7 upstream of the E1 sequence in HPV-18 [87]. This sequence includes several minicistrons and may form a secondary mRNA structure that is necessary for shunting. In contrast, Stacey et al. observed that E7 translation in HPV-16 likely occurred from leaky scanning of polycistronic messages [100,101]. Support for this stems from the observation that translation of E7 was found not to require splicing events of the upstream E6 ORF, and that E7 synthesis occurred from a mutant transcript in which the E6 ORF overlapped E7. It appears that ribosomes recognize the E6 AUG, yet ignore this start site as well as a number of AUGs in the E6 ORF to utilize the start codon for E7. Interestingly, E7 is not translated at a significant level from polycistronic E6-E7 transcripts in HPV-11 [100]. This low-risk virus expresses transcripts from a promoter at nt 229, where E7 is the first ORF on the message.

Conclusion

Papillomavirus gene expression has been intensely studied for the past two decades. The epitheliotropism of the virus and its synchronicity with the life-cycle of the infected cell pose two intriguing questions: what is the mechanism for establishing tissue preference, and how does the virus use host transcription factors to regulate its life-cycle. New insights into this process have come from studies involving the assembly of nucleoprotein complexes to regulate enhancer function. Additional insights come from studies of post-transcriptional mechanisms, including splicing, polyadenylation, and translation. Post-transcriptional mechanisms, in particular, provide a complex framework for the control of viral gene expression. A thorough understanding of HPV gene expression will, in the future, require an integrated view of transcription, RNA processing, and translation.

Acknowledgements

L. Peña is a student in the Medical Scientist Training Program and is supported by a grant from the National Cancer Institute (1F31CA80673-01). L.A. Laimins is funded by the National Institute for Allergy and Infectious Disease and by the National Cancer Institute.

References

1. Ai W, Narahari J, Roman A. Yin yang 1 negatively regulates the differentiation-specific E1 promoter of human papillomavirus type 6. J Virol 2000; 74: 5198–5205.
2. Ai W, Toussaint E, Roman A. CCAAT displacement protein binds to and negatively regulates human papillomavirus type 6 E6, E7, and E1 promoters. J Virol 1999; 73: 4220–4229.
3. Amin AA, Titolo S, Pelletier A, Fink D, Cordingley MG, Archambault J. Identification of domains of the HPV11 E1 protein required for DNA replication *in vitro*. Virology 2000; 272: 137–150.
4. Angel P, Szabowski A, Schorpp-Kistner M. Function and regulation of AP-1 subunits in skin physiology and pathology. Oncogene 2001; 20: 2413–2423.
5. Antson AA, Burns JE, Moroz OV, Scott DJ, Sanders CM, Bronstein IB, Dodson GG, Wilson KS, Maitland NJ. Structure of the intact transactivation domain of the human papillomavirus E2 protein. Nature 2000; 403: 805–809.
6. Apt D, Chong T, Liu Y, Bernard HU. Nuclear factor I and epithelial cell-specific transcription of human papillomavirus type 16. J Virol 1993; 67: 4455–4463.
7. Apt D, Watts RM, Suske G, Bernard HU. High Sp1/Sp3 ratios in epithelial cells during epithelial differentiation and cellular transformation correlate with the activation of the HPV-16 promoter. Virology 1996; 224: 281–291.
8. Bauknecht T, Angel P, Royer HD, zur Hausen H. Identification of a negative regulatory domain in the human papillomavirus type 18 promoter: interaction with the transcriptional repressor YY1. EMBO J 1992; 11: 4607–4617.

9. Bauknecht T, Jundt F, Herr I, Oehler T, Delius H, Shi Y, Angel P, zur Hausen H. A switch region determines the cell type-specific positive or negative action of YY1 on the activity of the human papillomavirus type 18 promoter. J Virol 1995; 69: 1–12.

10. Bedell MA, Hudson JB, Golub TR, Turyk ME, Hosken M, Wilbanks GD, Laimins LA. Amplification of human papillomavirus genomes *in vitro* is dependent on epithelial differentiation. J Virol 1991; 65: 2254–2260.

11. Bedrosian CL, Bastia D. The DNA-binding domain of HPV-16 E2 protein interaction with the viral enhancer: protein-induced DNA bending and role of the nonconserved core sequence in binding site affinity. Virology 1990; 174: 557–575.

12. Berg M, Stenlund A. Functional interactions between papillomavirus E1 and E2 proteins. J Virol 1997; 71: 3853–3863.

13. Berger SL, Felsenfeld G. Chromatin goes global. Mol Cell 2001; 8: 263–268.

14. Bouallaga I, Massicard S, Yaniv F, Thierry M. An enhanceosome containing the Jun B/Fra-2 heterodimer and the HMG-I(Y) architectural protein controls HPV 18 transcription. EMBO Rep 2000; 1: 422–427.

15. Burns JE, Moroz OV, Antson AA, Sanders CM, Wilson KS, Maitland NJ. Expression, crystallization and preliminary X-ray analysis of the E2 transactivation domain from papillomavirus type 16. Acta Crystallogr D Biol Crystallogr 1998; 54: 1471–1474.

16. Bussiere DE, Kong X, Egan DA, Walter K, Holzman TF, Lindh F, Robins T, Giranda VL. Structure of the E2 DNA-binding domain from human papillomavirus serotype 31 at 2.4 Å. Acta Crystallogr D Biol Crystallogr 1998; 54: 1367–1376.

17. Butz K, Hoppe-Seyler F. Transcriptional control of human papillomavirus (HPV) oncogene expression: composition of the HPV type 18 upstream regulatory region. J Virol 1993; 67: 6476–6486.

18. Chan WK, Klock G, Bernard HU. Progesterone and glucocorticoid response elements occur in the long control regions of several human papillomaviruses involved in anogenital neoplasia. J Virol 1989; 63: 3261–3269.

19. Chao SF, Rocque WJ, Daniel S, Czyzyk LE, Phelps WC, Alexander KA. Subunit affinities and stoichiometries of the human papillomavirus type 11 E1: E2: DNA complex. Biochemistry 1999; 38: 4586–4594.

20. Chiang CM, Dong G, Broker TR, Chow LT. Control of human papillomavirus type 11 origin of replication by the E2 family of transcription regulatory proteins. J Virol 1992; 66: 5224–5231.

21. Chiang CM, Ustav M, Stenlund A, Ho TF, Broker TR, Chow LT. Viral E1 and E2 proteins support replication of homologous and heterologous papillomaviral origins. Proc Natl Acad Sci USA 1992; 89: 5799–5803.

22. Chong T, Apt D, Gloss B, Isa M, Bernard HU. The enhancer of human papillomavirus type 16: binding sites for the ubiquitous transcription factors Oct-1, NFA, TEF-2, NF1, and AP-1 participate in epithelial cell-specific transcription. J Virol 1991; 65: 5933–5943.

23. Chong T, Chan WK, Bernard HU. Transcriptional activation of human papillomavirus 16 by nuclear factor I, AP1, steroid receptors and a possibly novel transcription factor, PVF: a model for the composition of genital papillomavirus enhancers. Nucleic Acids Res 1990; 18: 465–470.

24. Colgan DF, Manley JL. Mechanism and regulation of mRNA polyadenylation. Genes Dev 1997; 11: 2755–2766.

25. Conger KL, Liu JS, Kuo SR, Chow LT, Wang TS. Human papillomavirus DNA

replication. Interactions between the viral E1 protein and two subunits of human DNA polymerase alpha/primase. J Biol Chem 1999; 274: 2696–2705.

26. Cooper CS, Upmeyer SN, Winokur PL. Identification of single amino acids in the human papillomavirus 11 E2 protein critical for the transactivation or replication functions. Virology 1998; 241: 312–322.

27. del Mar Pena LM, Laimins LA. Differentiation-dependent chromatin rearrangement coincides with activation of human papillomavirus type 31 late gene expression. J Virol 2001; 75: 10005–10013.

28. Demeret C, Le Moal M, Yaniv M, Thierry F. Control of HPV 18 DNA replication by cellular and viral transcription factors. Nucleic Acids Res 1995; 23: 4777–4784.

29. Dietrich-Goetz W, Kennedy IM, Levins B, Stanley MA, Clements JB. A cellular 65-kDa protein recognizes the negative regulatory element of human papillomavirus late mRNA. Proc Natl Acad Sci USA 1997; 94: 163–168.

30. Dixon EP, Pahel GL, Rocque WJ, Barnes JA, Lobe DC, Hanlon MH, Alexander KA, Chao SF, Lindley K, Phelps WC. The E1 helicase of human papillomavirus type 11 binds to the origin of replication with low sequence specificity. Virology 2000; 270: 345–357.

31. Dollard SC, Broker TR, Chow LT. Regulation of the human papillomavirus type 11 E6 promoter by viral and host transcription factors in primary human keratinocytes. J Virol 1993; 67: 1721–1726.

32. Dong G, Broker TR, Chow LT. Human papillomavirus type 11 E2 proteins repress the homologous E6 promoter by interfering with the binding of host transcription factors to adjacent elements. J Virol 1994; 68: 1115–1127.

33. Doorbar J, Parton A, Hartley K, Banks L, Crook T, Stanley M, Crawford L. Detection of novel splicing patterns in a HPV16-containing keratinocyte cell line. Virology 1990. 178: 254–262.

34. Dostatni N, Thierry F, Yaniv M. A dimer of BPV-1 E2 containing a protease resistant core interacts with its DNA target. EMBO J 1988; 7: 3807–3816.

35. Edwalds-Gilbert G, Milcarek C. Regulation of poly(A) site use during mouse B-cell development involves a change in the binding of a general polyadenylation factor in a B-cell stage-specific manner. Mol Cell Biol. 1995; 15: 6420–6429.

36. Enemark EJ, Chen G, Vaughn DE, Stenlund A, Joshua-Tor L. Crystal structure of the DNA binding domain of the replication initiation protein E1 from papillomavirus. Mol Cell 2000; 6: 149–158.

37. Enzenauer C, Mengus G, Lavigne A, Davidson I, Pfister H, May M. Interaction of human papillomavirus 8 regulatory proteins E2, E6 and E7 with components of the TFIID complex. Intervirology 1998; 41: 80–90.

38. Ferreiro DU, Lima LM, Nadra AD, Alonso LG, Goldbaum FA, de Prat-Gay G. Distinctive cognate sequence discrimination, bound DNA conformation, and binding modes in the E2 C-terminal domains from prototype human and bovine papillomaviruses. Biochemistry 2000; 39: 14692–14701.

39. Flores ER, Allen-Hoffmann BL, Lee D, Sattler CA, Lambert PF. Establishment of the human papillomavirus type 16 (HPV-16) life cycle in an immortalized human foreskin keratinocyte cell line. Virology 1999; 262: 344–354.

40. Frattini MG, Laimins LA. The role of the E1 and E2 proteins in the replication of human papillomavirus type 31b. Virology 1994; 204: 799–804.

41. Frattini MG, Lim HB, Doorbar J, Laimins LA. Induction of human papillomavirus type

18 late gene expression and genomic amplification in organotypic cultures from transfected DNA templates. J Virol 1997; 71: 7068–7072.

42. Frattini MG, Lim HB, Laimins LA. *In vitro* synthesis of oncogenic human papillomaviruses requires episomal genomes for differentiation-dependent late expression. Proc Natl Acad Sci USA 1996; 93: 3062–3067.

43. Furth PA, Baker CC. An element in the bovine papillomavirus late 3′ untranslated region reduces polyadenylated cytoplasmic RNA levels. J Virol 1991; 65: 5806–5812.

44. Furth PA, Choe WT, Rex JH, Byrne JC, Baker CC. Sequences homologous to 5′ splice sites are required for the inhibitory activity of papillomavirus late 3′ untranslated regions. Mol Cell Biol 1994; 14: 5278–5289.

45. Gloss B, Yeo-Gloss M, Meisterenst M, Rogge L, Winnacker EL, Bernard HU. Clusters of nuclear factor I binding sites identify enhancers of several papillomaviruses but alone are not sufficient for enhancer function. Nucleic Acids Res 1989; 17: 3519–3533.

46. Gopalakrishnan V, Sheahan L, Khan SA. DNA replication specificity and functional E2 interaction of the E1 proteins of human papillomavirus types 1a and 18 are determined by their carboxyl-terminal halves. Virology 1999; 256: 330–339.

47. Grassmann K, Rapp B, Maschek H, Petry KU, Iftner T. Identification of a differentiation-inducible promoter in the E7 open reading frame of human papillomavirus type 16 (HPV-16) in raft cultures of a new cell line containing high copy numbers of episomal HPV-16 DNA. J Virol 1996; 70: 2339–2349.

48. Harris SF, Botchan MR. Crystal structure of the human papillomavirus type 18 E2 activation domain. Science 1999; 284: 1673–1677.

49. Hegde RS, Grossman SR, Laimins LA, Sigler PB. Crystal structure at 1.7 Å of the bovine papillomavirus-1 E2 DNA-binding domain bound to its DNA target. Nature 1992; 359: 505–512.

50. Higgins GD, Uzelin DM, Phillips GE, McEvoy P, Marin R, Burrell CJ. Transcription patterns of human papillomavirus type 16 in genital intraepithelial neoplasia: evidence for promoter usage within the E7 open reading frame during epithelial differentiation. J Gen Virol. 1992; 73: 2047–2057.

51. Hoppe-Seyler F, Butz K, zur Hausen H. Repression of the human papillomavirus type 18 enhancer by the cellular transcription factor Oct-1. J Virol 1991; 65: 5613–5618.

52. Hou SY, Wu SY, Zhou T, Thomas MC, Chiang CM. Alleviation of human papillomavirus E2-mediated transcriptional repression via formation of a TATA binding protein (or TFIID)-TFIIB-RNA polymerase II-TFIIF preinitiation complex. Mol Cell Biol 2000; 20: 113–125.

53. Howley PM. Papillomavirinae: the viruses and their replication. In: BN Fields, DM Knipe and PM Howley (Eds), Fundamental Virology, 3rd Edn. Lippincott-Raven, Philadelphia, 1996, pp. 947–978.

54. Hubert WG, Kanaya T, Laimins LA. DNA replication of human papillomavirus type 31 is modulated by elements of the upstream regulatory region that lie 5' of the minimal origin. J Virol 1999; 73: 1835–1845.

55. Hughes FJ, Romanos MA. E1 protein of human papillomavirus is a DNA helicase/ ATPase. Nucleic Acids Res 1993; 21: 5817–5823.

56. Hummel M, Hudson JB, Laimins LA. Differentiation-induced and constitutive transcription of human papillomavirus type 31b in cell lines containing viral episomes. J Virol 1992; 66: 6070–6080.

57. Hummel M, Lim HB, Laimins LA. Human papillomavirus type 31b late gene expression is regulated through protein kinase C-mediated changes in RNA processing. J Virol 1995; 69: 3381–3388.

58. Ishiji T, Lace MJ, Parkkinen S, Anderson RD, Haugen TH, Cripe TP, Xiao JH, Davidson I, Chambon P, Turek LP. Transcriptional enhancer factor (TEF)-1 and its cell-specific co-activator activate human papillomavirus-16 E6 and E7 oncogene transcription in keratinocytes and cervical carcinoma cells. EMBO J 1992; 11: 2271–2281.

59. Jenkins O, Earnshaw D, Sarginson G, Del Vecchio A, Tsai J, Kallender H, Amegadzie B, Browne M. Characterization of the helicase and ATPase activity of human papillomavirus type 6b E1 protein. J Gen Virol 1996; 77: 1805–1809.

60. Joyce JG, Tung JS, Przysiecki CT, Cook JC, Lehman ED, Sands JA, Jansen KU, Keller PM. The L1 major capsid protein of human papillomavirus type 11 recombinant virus-like particles interacts with heparin and cell-surface glycosaminoglycans on human keratinocytes. J Biol Chem 1999; 274: 5810–5822.

61. Karlen S, Offord EA, Beard P. Functional promoters in the genome of human papillomavirus type 6b. J Gen Virol 1996; 77: 11–16.

62. Kennedy IM, Haddow JK, Clements JB. A negative regulatory element in the human papillomavirus type 16 genome acts at the level of late mRNA stability. J Virol 1991; 65: 2093–2097.

63. Kleinheinz A, von Knebel Doeberitz M, Cripe TP, Turek LP, Gissmann L. Human papillomavirus early gene products and maintenance of the transformed state of cervical cancer cells *in vitro*. Curr Top Microbiol Immunol 1989. 144: 175–179.

64. Koffa MD, Graham SV, Takagaki Y, Manley JL, Clements JB. The human papillomavirus type 16 negative regulatory RNA element interacts with three proteins that act at different posttranscriptional levels. Proc Natl Acad Sci USA 2000; 97: 4677–4682.

65. Kyo S, Inoue M, Nishio Y, Nakanishi K, Akira S, Inoue H, Yutsudo M, Tanizawa O, Hakura A. NF-IL6 represses early gene expression of human papillomavirus type 16 through binding to the noncoding region. J Virol 1993; 67: 1058–1066.

66. Kyo S, Tam A, Laimins LA. Transcriptional activity of human papillomavirus type 31b enhancer is regulated through synergistic interaction of AP1 with two novel cellular factors. Virology 1995; 211: 184–197.

67. Lai MC, Teh BH, Tarn WY. A human papillomavirus E2 transcriptional activator. The interactions with cellular splicing factors and potential function in pre-mRNA processing. J Biol Chem 1999; 274: 11832–11841.

68. Laimins LA. The biology of human papillomaviruses: from warts to cancer. Infect Agents Dis 1993. 2: 74–86.

69. Lee D, Sohn H, Kalpana GV, Choe J. Interaction of E1 and hSNF5 proteins stimulates replication of human papillomavirus DNA. Nature 1999; 399: 487–491.

70. Lee KY, Broker TR, Chow LT. Transcription factor YY1 represses cell-free replication from human papillomavirus origins. J Virol 1998; 72: 4911–4917.

71. Maniatis T, Falvo JV, Kim TH, Kim TK, Lin CH, Parekh BS, Wathelet MG. Structure and function of the interferon-B enhanceosome. Cold Spring Harb Symp Quant Biol 1998; 63: 609–620.

72. Meyers C, Frattini MG, Hudson JB, Laimins LA. Biosynthesis of human papillomavirus from a continuous cell line upon epithelial differentiation. Science 1992; 257: 971–973.

73. Meyers C, Laimins LA. *In vitro* systems for the study and propagation of human

papillomaviruses. Curr Top Microbiol Immunol 1994; 186: 199–215.

74. Meyers C, Mayer TJ, Ozbun MA. Synthesis of infectious human papillomavirus type 18 in differentiating epithelium transfected with viral DNA. J Virol 1997; 71: 7381–7386.

75. Mittal R, Pater A, Pater MM. Multiple human papillomavirus type 16 glucocorticoid response elements functional for transformation, transient expression, and DNA–protein interactions. J Virol 1993; 67: 5656–5659.

76. Morris PJ, Dent CL, Ring CJ, Latchman DS. The octamer binding site in the HPV16 regulatory region produces opposite effects on gene expression in cervical and non-cervical cells. Nucleic Acids Res 1993; 21: 1019–1023.

77. Nepveu A. Role of the multifunctional CDP/Cut/Cux homeodomain transcription factor in regulating differentiation, cell growth and development. Gene 2001; 270: 1–15.

78. Nilsson CH, Bakos E, Petry KU, Schneider A, Durst M. Promoter usage in the E7 ORF of HPV16 correlates with epithelial differentiation and is largely confined to low-grade genital neoplasia. Int J Cancer 1996; 65: 6–12.

79. O'Connor M, Bernard HU. Oct-1 activates the epithelial-specific enhancer of human papillomavirus type 16 via a synergistic interaction with NFI at a conserved composite regulatory element. Virology 1995; 207: 77–88.

80. O'Connor MJ, Stunkel W, Koh CH, Zimmermann H, Bernard HU. The differentiation-specific factor CDP/Cut represses transcription and replication of human papilloma-viruses through a conserved silencing element. J Virol 2000; 74: 401–410.

81. O'Connor MJ, Tan SH, Tan CH, Bernard HU. YY1 represses human papillomavirus type 16 transcription by quenching AP-1 activity. J Virol 1996; 70: 6529–6539.

82. Ozbun MA, Meyers C. Characterization of late gene transcripts expressed during vegetative replication of human papillomavirus type 31b. J Virol 1997; 71: 5161–5172.

83. Ozbun MA, Meyers C. Temporal usage of multiple promoters during the life cycle of human papillomavirus type 31b. J Virol 1998; 72: 2715–2722.

84. Pattison S, Skalnik DG, Roman A. CCAAT displacement protein, a regulator of differ-entiation-specific gene expression, binds a negative regulatory element within the 5′ end of the human papillomavirus type 6 long control region. J Virol 1997; 71: 2013–2022.

85. Pray TR, Laimins LA. Differentiation-dependent expression of E1–E4 proteins in cell lines maintaining episomes of human papillomavirus type 31b. Virology 1995; 206: 679–685.

86. Rapp B, Pawellek A, Kraetzer F, Schaefer M, May C, Purdie K, Grassmann K, Iftner T. Cell-type-specific separate regulation of the E6 and E7 promoters of human papilloma-virus type 6a by the viral transcription factor E2. J Virol 1997; 71: 6956–6966.

87. Remm M, Remm A, Ustav M. Human papillomavirus type 18 E1 protein is translated from polycistronic mRNA by a discontinuous scanning mechanism. J Virol 1999; 73: 3062–3070.

88. Rohlfs M, Winkenbach S, Meyer S, Rupp T, Durst M. Viral transcription in human keratinocyte cell lines immortalized by human papillomavirus type-16. Virology 1991; 183: 331–342.

89. Rotenberg MO, Chow LT, Broker TR. Characterization of rare human papillomavirus type 11 mRNAs coding for regulatory and structural proteins, using the polymerase chain reaction. Virology 1989; 172: 489–497.

90. Rudge TL, Johnson LF. Inactivation of MED-1 elements in the TATA-less, initiator-less mouse thymidylate synthase promoter has no effect on promoter strength or the complex

pattern of transcriptional start sites. J Cell Biochem 1999; 73: 90–96.

91. Ruesch MN, Stubenrauch F, Laimins LA. Activation of papillomavirus late gene transcription and genome amplification upon differentiation in semisolid medium is coincident with expression of involucrin and transglutaminase but not keratin-10. J Virol 1998; 72: 5016–5024.

92. Sakai H, Yasugi T, Benson JD, Dowhanick JJ, Howley PM. Targeted mutagenesis of the human papillomavirus type 16 E2 transactivation domain reveals separable transcriptional activation and DNA replication functions. J Virol 1996; 70: 1602–1611.

93. Sang BC, Barbosa MS. Increased E6/E7 transcription in HPV 18-immortalized human keratinocytes results from inactivation of E2 and additional cellular events. Virology 1992; 189: 448–455.

94. Seedorf K, Oltersdorf T, Krammer G, Rowekamp W. Identification of early proteins of the human papilloma viruses type 16 (HPV 16) and type 18 (HPV 18) in cervical carcinoma cells. EMBO J 1987. 6: 139–144.

95. Sibbet GJ, Cuthill S, Campo MS. The enhancer in the long control region of human papillomavirus type 16 is up-regulated by PEF-1 and down-regulated by Oct-1. J Virol 1995; 69: 4006–5011.

96. Smotkin D, Prokoph H, Wettstein FO. Oncogenic and nononcogenic human genital papillomaviruses generate the E7 mRNA by different mechanisms. J Virol 1989. 63: 1441–1447.

97. Snijders PJ, van den Brule AJ, Schrijnemakers HF, Raaphorst PM, Meijer CJ, Walboomers JM. Human papillomavirus type 33 in a tonsillar carcinoma generates its putative E7 mRNA via two E6* transcript species which are terminated at different early region poly(A) sites. J Virol 1992; 66: 3172–3178.

98. Sokolowski M, Schwartz S. Heterogeneous nuclear ribonucleoprotein C binds exclusively to the functionally important UUUUU-motifs in the human papillomavirus type-1 AU-rich inhibitory element. Virus Res 2001; 73: 163–175.

99. Sokolowski M, Tan W, Jellne M, Schwartz S. mRNA instability elements in the human papillomavirus type 16 L2 coding region. J Virol 1998; 72: 1504–1515.

100. Stacey SN, Jordan D, Snijders PJ, Mackett M, Walboomers JM, Arrand JR. Translation of the human papillomavirus type 16 E7 oncoprotein from bicistronic mRNA is independent of splicing events within the E6 open reading frame. J Virol 1995; 69: 7023–7031.

101. Stacey SN, Jordan D, Williamson AJ, Brown M, Coote JH, Arrand JR. Leaky scanning is the predominant mechanism for translation of human papillomavirus type 16 E7 oncoprotein from E6/E7 bicistronic mRNA. J Virol 2000; 74: 7284–7297.

102. Steger G, Corbach S. Dose-dependent regulation of the early promoter of human papillomavirus type 18 by the viral E2 protein. J Virol 1997; 71: 50–58.

103. Stubenrauch F, Colbert AM, Laimins LA. Transactivation by the E2 protein of oncogenic human papillomavirus type 31 is not essential for early and late viral functions. J Virol 1998; 72: 8115–8123.

104. Stubenrauch F, Hummel M, Iftner T, Laimins LA. The E8 ^ E2C protein, a negative regulator of viral transcription and replication, is required for extrachromosomal maintenance of human papillomavirus type 31 in keratinocytes. J Virol 2000; 74: 1178–1186.

105. Stubenrauch F, Zobel T, Iftner T. The E8 domain confers a novel long-distance transcriptional repression activity on the E8 ^ E2C protein of high-risk human papillomavirus type 31. J Virol 2001; 75: 4139–4149.

106. Stunkel W, Bernard HU. The chromatin structure of the long control region of human papillomavirus type 16 represses viral oncoprotein expression. J Virol 1999; 73: 1918–1930.

107. Tan SH, Gloss B, Bernard HU. During negative regulation of the human papillomavirus-16 E6 promoter, the viral E2 protein can displace Sp1 from a proximal promoter element. Nucleic Acids Res 1992; 20: 251–256.

108. Terhune SS, Hubert WG, Thomas JT, Laimins LA. Early polyadenylation signals of human papillomavirus type 31 negatively regulate capsid gene expression. J Virol 2001; 75: 8147–8157.

109. Terhune SS, Milcarek C, Laimins LA. Regulation of human papillomavirus type 31 polyadenylation during the differentiation-dependent life cycle. J Virol 1999; 73: 7185–7192.

110. Tomita Y, Shiga T, Simizu B. Characterization of a promoter in the E7 open reading frame of human papillomavirus type 11. Virology 1996; 225: 267–273.

111. Wang H, Liu K, Yuan F, Berdichevsky L, Taichman LB, Auborn K. C/EBPbeta is a negative regulator of human papillomavirus type 11 in keratinocytes. J Virol 1996; 70: 4839–4844.

112. Yukawa K, Butz K, Yasui T, Kikutani H, Hoppe-Seyler F. Regulation of human papillomavirus transcription by the differentiation- dependent epithelial factor Epoc-1/skn-1a. J Virol 1996; 70: 10–16.

113. Zhao W, Chow LT, Broker TR. A distal element in the HPV-11 upstream regulatory region contributes to promoter repression in basal keratinocytes in squamous epithelium. Virology 1999; 253: 219–229.

114. Zhao W, Chow LT, Broker TR. Transcription activities of human papillomavirus type 11 E6 promoter—proximal elements in raft and submerged cultures of foreskin keratinocytes. J Virol 1997; 71: 8832–8840.

Human Papillomaviruses
D.J. McCance (editor)

Human papillomavirus DNA replication

Jen-Sing Liu and Thomas Melendy

Department of Microbiology, University at Buffalo, State University of New York, School of Medicine &
Biomedical Sciences, 213 Biomedical Research Building, 3435 Main Street, Buffalo, New York 14214, USA

Introduction

Human papillomaviruses (HPV) are a family of double-stranded circular DNA viruses with a genomic size of around 8 kilobases. HPV are highly species-specific and infect only stratified squamous epithelial cells or mucosal membranes [117]. As with many other aspects of HPV molecular biology, the DNA replication of bovine papillomavirus type 1 (BPV) has been the most studied and has become the paradigm for HPV DNA replication. DNA replication studies with various HPVs have shown that although the BPV paradigm is often accurate, differences do exist between the viral DNA replication of BPV and some HPVs.

Viruses infect at the basal layer of epidermal tissue and the virus undergoes three stages of replication. Within the basal and first supra-basal epidermal layers, the viral DNA is propagated until there are 50–100 copies per cell. During the second stage, the maintenance stage, viral DNA replication is replicated in synchrony with the host cell cycle, only occurring during S phase of the host cell cycle [32]. During both of these stages the HPV DNA is maintained as episomal plasmids in the nuclei of infected cells. The third and final stage is the vegetative stage of viral replication. The vegetative stage only occurs in terminally differentiated tissues, and includes both an increase in the number of viral genome copies, as well as the expression of late genes and assembly of new virus [13]. The requirement for cellular differentiation for the complete virus life-cycle has greatly hampered study of papillomavirus DNA replication [14,91].

Model systems

Our current knowledge of the molecular mechanisms of papillomavirus (PV) DNA replication comes primarily from model systems. One critical advancement was the development of a transient transfection system for the study of PV DNA replication. In this system cultured cells are transfected with two plasmids: one is the expression plasmid, which is used to express PV proteins required for DNA replication; and the other is the test plasmid, which is used to test for *cis*-acting DNA elements required for DNA replication. Using this system both the PV proteins and sequences required to drive episomal PV DNA replication were identified [104–106]. The origin of PV DNA replication is located within the upstream regulatory region (URR) or long

54

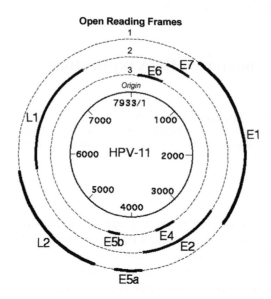

Fig. 1. The structure of the HPV-11 genome. The solid circle represents the 7933 base pair dsDNA genome. Nucleotide positions within the genome are indicated by the numbers inside the solid circle. The encoded proteins (ORFs) are indicated by the thick solid lines in the outer circles. The URR/LCR is in the open region between ORFs L1 and E6. The location of the origin is indicated within the Early gene proximal side of the URR/LCR.

control region (LCR) within the major non-coding region of the HPV genome (Fig. 1). Only two virally encoded proteins, E1 and E2, are required for the replication of PV episomal DNA. All other factors required for PV DNA replication are obtained from the host cell.

Eukaryotic cell extracts supplemented with the appropriate PV E1 and E2 proteins can be used to reconstitute replication of DNA episomes containing both bovine and human PV origin sequence in a cell-free *in vitro* system [44,108]. This system has been compared to the well-studied simian virus 40 (SV40) DNA replication *in vitro* system. All the host factors required for SV40 DNA replication are also required for PV DNA replication [44,49,63,65,69,108]; however, they are not sufficient [63]. Additional cellular factors are required for PV DNA replication. In further support of this, it was found that extracts from some types of human cells that support SV40 DNA replication efficiently, are deficient for PV DNA replication [50].

Another model system is the organotypic or raft culture system [68]. As with the transient transfection system above, plasmids are introduced into the cells to be cultured. However, in this system the plasmids used are the entire viral genomic DNA. These cells are then grown on a collagen raft at the media–air interface. This system more closely mimics the natural viral life-cycle, as these cells grown at a liquid–air interface are induced to undergo differentiation. This system has also been

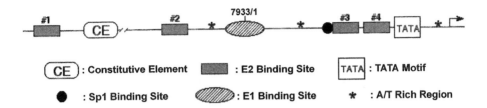

Fig. 2. Organization of the HPV origin. The solid line represents the dsDNA of the HPV-11 origin region from nucleotide 7500 to nucleotide 100 (as noted by the designated numbers above the line). The E1, E2, and Sp1 binding site positions are indicated by the striped oval, gray rectangles, and solid circle as indicated. The E2 binding sites within this region are numbered in bold. The TATA box indicates the TATA element for the major E6 promoter (start site indicated by the arrow), and the Constitutive Element is indicated by the rounded rectangle. The A/T rich regions are indicated by asterisks.

used to study the *cis*-acting DNA elements and viral *trans*-acting factors required for HPV origin DNA replication [67,92,93]. The limitation with this system is that many aspects of PV biology are required for propagation of viral DNA. Hence, if a studied factor is involved in several functional processes (for example, E2 is involved in both viral transcription and DNA replication regulation), elucidation of its role in PV DNA replication may be difficult.

HPV Replication origins

As noted above, the origin of PV DNA replication is located within the untranslated upstream regulatory region (URR) or long control region (LCR) and overlaps the major E6 promoter (Fig. 2). The *cis*-element composition of HPV DNA replication origins is relatively simple. The core origin is located within a 200 base pair (bp) sequence containing one E1 protein binding sequence (E1BS), three important E2 protein binding sequences (E2BS) (sites 2, 3 and 4) and surrounding sequences.

E1 protein binding sequence (E1BS)

The E1BS was first identified in BPV as an 18 base-pair imperfect palindromic sequence [41]. This AT-rich sequence is highly conserved among PVs (Fig. 3) [14]. Based on studies of the interaction between BPV E1 and its E1BS, it was suggested that there are four E1 protein binding sites within the 18 bp E1BS (Fig. 3) [9,26]. A similar structure has also been found in the simian virus 40 (SV40) origin of DNA replication. Binding Site II, the essential central element of the SV40 core origin, consists of four binding sites for the SV40 large T antigen (Tag) [6].

While an E1BS is essential for function for both the BPV and HPV type 1 origins [36,89,105,110], it is dispensable for origin activity in both transient replication assays and cell-free DNA replication for many HPV isotypes [21,54,77,95]. However, even when not essential, the presence of an E1BS can stimulate origin activity dramatically [77].

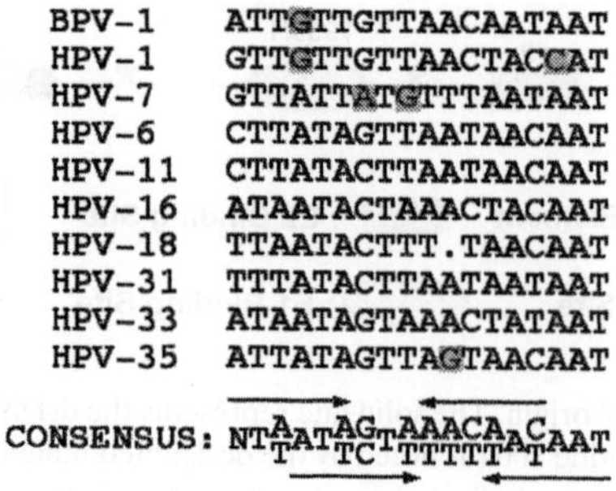

```
BPV-1     ATTGTTGTTAACAATAAT
HPV-1     GTTGTTGTTAACTACCAT
HPV-7     GTTATTATGTTTAATAAT
HPV-6     CTTATAGTTAATAACAAT
HPV-11    CTTATACTTAATAACAAT
HPV-16    ATAATACTAAACTACAAT
HPV-18    TTAATACTTT.TAACAAT
HPV-31    TTTATACTTAATAATAAT
HPV-33    ATAATAGTAAACTATAAT
HPV-35    ATTATAGTTAGTAACAAT
```

CONSENSUS: NTₜAₐTAGₜTAAACAₐCₐₐT (with variant bases below)

Fig. 3. Homology of PV E1BSs. The sequences of the indicated E1BSs are displayed over the E1BS consensus sequence. Bases which do not adhere to the consensus are grayed. The arrows indicate the putative individual E1 binding sites, with the arrowhead indicating the relative orientation [26].

E2 protein binding sequences (E2BS)

For many HPV isotypes, the only *cis*-element required for replication function in transient transfection assays is the E2BS ($ACCGN_4CGGT$) (HPV-1 is the only exception, see Ref. [38]). A single copy of either a natural or synthetic E2BS is both required and sufficient for HPV origin function in these assays [54,77,95]. The genomes of most HPV isotypes contain four copies of the E2BS within the URR (Fig. 2). The contribution of each E2BS to origin activity is based on its relative proximity to the E1BS and the affinity of the E2 protein for that specific site [77]. Origin activity also increases with the number of E2BSs. The three most proximal E2BSs (#2 to #4) plus the E1BS provides full origin activity in transient transfection, *in vitro* cell-free, and raft culture DNA replication assays [14,91].

Auxiliary sequences

Other than the E1BS and E2BSs, several *cis*-elements in the proximal region have also been examined for their possible contribution to the overall efficiency of origin function. These sequences are: the A/T-rich sequences surrounding the E1BS and E2BSs, a purine-rich sequence, and several transcription factor binding sites (Sp1, GT-1, AP-1, and a TATA motif) [12,20,22,38,54,77]. The results indicated that these sequences make little or no contribution to HPV origin activity.

In summary, the *cis*-elements important for HPV DNA replication origin function consist of an E1BS and the three surrounding E2BSs (Fig. 2).

Replication machinery

E1 and E2 are the only two virally encoded proteins required for PV DNA replication. All the other factors directly involved in PV DNA replication are provided by the host cell.

E1 protein

The E1 proteins of PVs all show a very high degree of conservation. They also have a limited degree of homology to the large T antigens (Tag) of SV40 and other polyomaviruses [15]. PV E1 proteins have both ATPase and DNA helicase activities that are essential for both initiation and elongation of viral DNA replication [14,91]. E1 interacts with the viral E2 protein, as well as several host factors. Many of these interactions have either been shown to or are suspected to be essential for PV DNA replication. E1 is a nuclear phosphoprotein that has low affinity for both single stranded DNA (ssDNA) and double-stranded DNA (dsDNA). The carboxy-terminal 2/3 of E1 contains most of the functional domains of E1 (Fig. 4) and contains the regions of homology between the various E1 proteins and SV40 Tag. Like SV40 Tag, PV E1 proteins bind to their respective origins, form hexameric ATPase/helicases, and are considered to be the proteins that initiate viral DNA replication.

   (i) *ATPase/helicase*: The ATPase and helicase domain is located within the C-terminal region of E1 [59,107] and shows a high degree of sequence homology to the superfamily 3 helicases (Fig. 4, Boxes A-D) [15]. MacPherson et al. [57] suggested that Box A is not required for the ATPase and helicase activities of HPV-16 E1; however, point mutants in Box A of HPV-11 E1 show substantially reduced ATPase

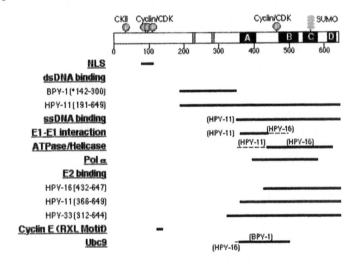

Fig. 4. Structure of the PV E1 protein. The structure of a consensus E1 protein is depicted at the top of the figure. Sites of phosphorylation on E1 by the indicated kinases are depicted by the circled P's. The site of sumoylation is similarly indicated. Hydrophilic regions 1 and 3 (HR1 and HR3) near the N-terminus are labeled as gray boxes. The dark boxes labeled A through D are the regions of highest homology between the E1 proteins, that show homology with SV40 T-antigen. These four domains also contain the major ATPase/helicase consensus sequence motifs. The regions mapped to be responsible for nuclear localization (NLS), dsDNA binding, ssDNA binding, E1–E1 interaction, ATPase/helicase function, DNA polymerase alpha binding, E2 binding, Cyclin E binding, and Ubc9 binding are indicated. *Based on the sequence alignment, the core DNA binding domain of BPV-1 E1 (amino acids 142-300) is located between amino acids 200–350 of HPV E1.

activity [102]. Mutations in Box B (which contains the Walker A and Walker B motifs, both of which are essential for ATP hydrolysis) also decrease E1 ATPase activity [57,75]. PV E1 ATPase activities have been shown to have K_m values within the physiological ATP concentration range [107], and are comparable to the ATPase activity of SV40 Tag. Like Tag, E1 forms a hexameric helicase that translocates in the 3′ to 5′ direction on the strand to which it remains bound [31,85,87,109].

(ii) *DNA binding*: While SV40 Tag and BPV E1 both show relatively high sequence specificity and affinity for dsDNA binding [27,91], most HPV E1 proteins bind dsDNA with low sequence-specificity and low affinity [14,21,51]. This may explain why the E1BS is not essential for origin activity for many HPVs. However, in the presence of an E2BS and E2 protein, the non-specific DNA binding activity of E1 is suppressed [5,44,108] and its sequence-specific DNA affinity is increased [4,7,14, 80,86,91].

The DNA binding properties of BPV E1 have been studied extensively. BPV E1 forms stable monomers (only in the presence of E2), dimers, trimers, tetramers, hexamers and double hexamers on DNA [8,10,26,31,84,85]. Binding of dsDNA by BPV E1 is ATP-independent [31,40,83], and a polypeptide of the BPV E1 protein spanning amino acid residues 142–300 (equivalent to amino acid residues 200–350 of HPV E1) carries the same sequence-specific DNA binding activity as the full length protein [8,35,46,82]. This domain is defined as the core DNA binding domain (DBD). The crystal structure of the BPV E1 DBD has been solved [26] and combined with biochemical studies shows that BPV E1 has a bipartite DBD with two clusters of hydrophilic amino acids (HR1 and HR3) [26,35]. These two regions are conserved among all PV E1 proteins [35].

While there is clear conservation of the DBD between BPV E1 and the HPV E1 proteins, the corresponding HPV E1 domain (amino acid residues 200–350) is unable to bind dsDNA. It appears that additional C-terminal sequences containing the conserved Boxes A–D, that carry ssDNA binding, ATP binding, ATPase, DNA helicase activity, and the E1 oligomerization domain, are required for dsDNA binding of the HPV E1 proteins [1,59,94,102,103,107]. For instance, a small deletion at the end of the C-terminus of HPV-11 E1 abolishes the DNA binding activity [94]. Furthermore, studies with HPV-11 E1 show that mutation of a highly conserved amino acid within Box A affects the dsDNA binding of E1, the E1–E1 interaction, and ATPase activities, leading to loss of the ability of E1 to support viral DNA replication [1,102,103,107]. Mutations in Box B also reduced the ATP and DNA binding ability of E1 [102]. Consistent with the mutational studies, formation of stable HPV E1–DNA complexes at physiological temperature require the presence of ATP [51,52,102]. While BPV E1 binds to DNA in various multimeric forms, HPV E1 is only found to bind as a hexamer. (In the presence of heat shock protein 40 (Hsp40) it is assembled into double hexamers [52]).

Based on both the BPV and HPV results, we hypothesize that when E1 first associates with the E1BS, it binds as dimers. This step does not require ATP. While this E1 dimer–DNA complex is relatively stable for BPV, it appears to be unstable for HPV. In the presence of ATP, more E1 proteins are recruited to the origin and assembled into E1 hexamers or double hexamers. Based on current models of hexameric DNA helicases, we propose that these E1 hexamers are formed such that they encircle one strand of the dsDNA helix, causing localized melting of the dsDNA in the process. These ssDNA-bound E1 hexamers appear to be stable. This model is consistent with the data showing that the BPV E1 DBD can bind dsDNA stably in the absence of ATP, while binding of dsDNA by HPV E1 requires ATP, and the ATPase/DNA helicase, and E1 oligomerization domains of HPV E1.

(iii) *E1–E2 interaction*: The interaction between E1 and E2 is required for HPV DNA replication [14,91], and is a possible target for therapeutic intervention. However, investigations to identify the domain of E1 that interacts with E2 have produced inconsistent results. The most recent studies on HPV E1 all indicate that the E2-interacting domain is within the C-terminal region of E1 (Fig. 4) [1,37,57,64,94, 103,111,115]. Although this domain also carries other functions, these functions can be separated from E2 binding. Several point mutations in Box A or B that affect the ATPase and DNA binding activities of E1, still bind E2 efficiently [1,102,103,107]. Interestingly, binding and hydrolysis of ATP by E1 reduces the affinity between E1 and E2 [80,107]. During the initiation of HPV DNA replication, the binding of E2 to the E2BS, and the E1–E2 interaction, act to bring the initiator protein, E1, to the origin [34,81]. After the assembly of the dihexameric E1 helicase on the origin, E1 and E2 dissociate from each other, allowing the helicase to progress from the origin sequence. Presumably either binding of ATP by E1, or interactions with host cell factors, help contribute to the E1–E2 dissociation.

(iv) *Post-translational modifications*: Both SV40 Tag and E1 are phosphoproteins. Phosphorylation of Tag on the threonine at amino acid residue 124, near the nuclear localization signal (NLS), by the $p34^{cdc2}$ kinase is essential for SV40 DNA replication (for review, see Ref. [72]). HPV E1 has a bipartite NLS located between amino acids 80 and 130 and contain several conserved basic amino acid clusters [47]. Within this domain, different E1s contain two or three serines each followed by a proline. These serine-proline sequences are minimal Cyclin/CDK consensus sequences, (S/T)P. It has been shown that Cyclin E/CDK can phosphorylate these sites on HPV-11 E1 *in vitro* [56] and mutation of these to alanines reduces the DNA replication activity of HPV-11 E1 [50,56]. E1 carries a minimal Cyclin-binding motif (RXL) at a conserved basic cluster region near the N-terminus (Fig. 4), and has been shown to interact with Cyclin E [19,50,56]. A serine followed by several acidic amino acids within the N-terminal region of E1 can be phosphorylated by casein kinase II (CKII) *in vitro* [61,62]. This serine is conserved among all PV E1 proteins and mutation to an alanine in BPV E1 has little effect on most E1 activities, but fails to support BPV origin-dependent replication in the transient transfection assay [61,62].

E1 was also found to be sumoylated in cells, apparently due to its interaction with the ubiquitin-conjugating E2 enzyme, Ubc9, shown through a yeast two-hybrid study [73,74,112,113]. Sumoylation sites on BPV E1 were mapped to lysine 514, which is conserved among many HPV E1s. An E1 mutant, lysine 514 to alanine (K514A), was shown to not undergo nuclear translocation; and was therefore not able to support PV DNA replication in transient transfected cells [73,74].

(v) *Interactions with host proteins*: In addition to the cellular proteins that interact with E1 to modify it, E1 also directly interacts with several other host proteins. As noted above, interaction with Hsp40 enhances the formation of E1 double hexamers on origin containing DNA [52]. Like SV40 Tag, E1 has been shown to interact with the cellular replication protein complexes, DNA polymerase alpha-primase (pol-prim) and replication protein A (RPA). The 180 and 68 kDa subunits of polprim have both been shown to interact with both Tag and E1 [1,16,17,23–25,59,69,88]. The polprim interacting domain on E1 partially overlaps with the E2 binding domain and binding of E2 and polprim to E1 are mutually exclusive events [1,17,59]. RPA is the predominant eukaryotic single-stranded DNA binding protein and SV40 Tag and E1 both interact with RPA through its major 70 kDa ssDNA binding subunit [39]. All these interactions appear to be important for initiation and elongation of viral DNA replication.

E1 has been used in the yeast two hybrid system to search for other E1 interacting proteins. In addition to the Ubc9 protein (see above) this assay also identified the human SWI/SNF chromatin remodeling family protein, hSNF5, as a putative E1 interacting protein [45]. Whether this interaction is involved in PV DNA replication remains unknown. E1 has also been shown to interact with histone H1 [96] and this may be related to the interaction with hSNF5.

E2 protein family

The full-length E2 protein plays an important role in both viral DNA replication and transcription modulation (for a review, see Ref. [90]). The E2 protein has three major domains: the N-terminal transcriptional transactivation domain, the non-structured hinge region, and the C-terminal dimerization/DNA binding domain. *In vivo*, alternative splicing of the E2 open reading frame produces two shorter proteins that carry the hinge region and the dimerization/DNA binding domain (Fig. 5). Both shorter proteins are transcription inhibitors and are unable to support PV DNA replication in place of full length E2 [12,51].

E2 proteins are sequence-specific DNA binding proteins. For HPVs, the E2BSs have a consensus sequence of $ACCGN_4CGGT$, where the four central nucleotides are A/T-rich [79,99]. E2 binds to the E2BS as a dimer; and the two molecules interact through their dimerization/DNA binding domains. Based on structural studies of the N-terminal domain of E2, it has been proposed that the two dimeric molecules can also interact through alpha helices 2 and 3 in the N-terminus [2]. Although it has been shown that E2 also shows high affinity for the sequence $ACACN_5GGT$ [66], physiological significance of this interaction has not been shown.

The major function of E2 in HPV DNA replication is to bring E1 to the origin through protein–protein and protein–DNA interactions. Since the E2BS is the only essential *cis*-element for HPV DNA replication, the E2 protein must be considered the primary origin recognition protein for HPV (HPV-1 is the only known exception, since for HPV-1 DNA replication, E1 has been shown to be the only required viral protein [36]). In transiently transfected cells, the hinge region of E2 attaches to the

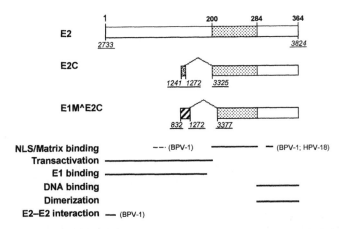

Fig. 5. Structure of the PV E2 protein family. The structure of a consensus E2 protein and the two shorter forms of E2 are depicted at the top of the figure. The three domains are indicated by the division of E2 into three sections. The N terminal 200 amino acid residues contain the transactivation and E1 binding regions. The N terminal domain also contains the E2–E2 interaction domain, at the far N-terminus. The nuclear matrix binding domain is within the central (hinge) region (amino acids 200 to 284, gray rectangle). The DNA binding domain is within the C-terminal domain (amino acids 284 to 364). In BPV, a secondary E1-E2 interaction has been noted between their respective DNA binding domains.

nuclear matrix [115], which are sites for host cell DNA replication. E2 is also capable of generating a nucleosome-free area at the PV origin and this is of particular importance as histone binding across the origin would otherwise inhibit replication activity [48].

The initial domain of E2 responsible for targeting E1 to the origin is the N-terminal transcriptional transactivation domain of E2. However, the transcriptional transactivation and E1 binding functions are separable. E2 point mutants at a highly conserved glutamine (amino acid residue 39 of BPV E2) lose their ability to interact with E1 and support PV DNA replication, but have no effect on the transactivation function of E2 [18,28,43,78]. Addition of a synthetic polypeptide of the same sequence as amino acid residues 33 to 45 of HPV-16 E2 is reported to block the E1-E2 interaction and HPV DNA replication [43].

The replication-related functions of the three domains of E2 clearly describe the roles of E2 in the initiation of PV DNA replication. The C-terminal DNA binding domain of E2 binds to the viral DNA, and the adjacent hinge region may act to target the viral DNA to sites of active DNA replication at the nuclear matrix [97,114]. The N-terminal domain of E2 can target the essential E1 initiator/helicase to the viral origin, and simultaneously to sites of active DNA replication [114]. Once viral DNA replication is initiated and the E1 helicase translocates away from the origin sequences, E2 ceases to be required [33,53,55,80].

E2 protein levels were found to be regulated throughout the cell cycle [110]. One of the mechanisms of regulation is through proteasome-mediated ubiquitination-targeted degradation. For BPV E2, phosphorylation of serine 301 in the hinge region is required for subsequent ubiquitination [70]. However, serine 301 is not conserved among the HPV E2 proteins and in fact HPV-18 E2 was found to be ubiquitinated at unknown amino acids in the N-terminus [3]. The hinge region was not required for this modification

Host factors

The *in vitro* cell free PV DNA replication assay has been used to identify the cellular proteins required for PV DNA replication. The host cell DNA replication factors required and sufficient for efficient SV40 DNA replication are: DNA polymerase alpha-primase and DNA polymerase delta, RPA, replication factor C, proliferating cell nuclear antigen (PCNA), topoisomerases I and II, RNase H, flap endonuclease (FEN-1) and DNA ligase I. Most of these factors have been shown to be required for BPV DNA replication *in vitro* [49,63,65,69], and it is presumed that RNase H, FEN-1 and DNA ligase I are required as well. We have shown that HPV DNA replication requires the same host proteins as BPV DNA replication (Shu-Ru Kuo, J.-S.L., and T.M., unpublished results). However, unlike with SV40, these factors are not sufficient to support appreciable levels of DNA synthesis [63]; and additional host factors are required for efficient PV DNA replication. These as yet unidentified factors are required in addition to those factors described above, and to Cyclin E, hSNF5, Hsp70 and Hsp40, which act to assist E1 and E2 in forming an active replication complex at the PV origin. The identity of these additional factors, and their role in PV DNA replication is still under investigation.

HPV DNA replication in terminally differentiated cells

HPV infection is highly species- and tissue-specific. Vegetative viral DNA replication can only be seen in terminally differentiated cells. However, with overexpressed E1 and E2, replication of HPV origin-containing plasmids can be studied in cycling human or animal cells. This indicates that the functional interactions between E1, E2 and the cellular replication proteins are neither cell type-specific nor differentiation-dependent. However, expression of E1 and E2 is regulated by differentiation (see Chapter 3). It has been shown that both E1 and E2 expression levels parallel viral DNA copy number across differentiated layers of cells in raft cultures [67]. Therefore, the differentiation dependence of HPV DNA replication appears to be due to transcriptional regulation of E1 and E2.

The viral E6 and E7 gene products also play critical roles for PV DNA replication in differentiated cells. Terminally differentiated cells are generally growth arrested (in G0). In G0 cells, the host cell DNA replication machinery is down regulated, and therefore not available for viral DNA replication. In order to promote DNA replication in these upper epidermal layers, the E6 and E7 proteins are required to drive the host cells into a pseudo-S phase (see Chapters 4 and 5). E6 binds to and inactivates p53, which prevents cellular checkpoint mechanisms from responding to unscheduled DNA replication [42,58,98,101,116]. Binding of E7 to Rb family proteins releases and activates the transcriptional transactivator E2F, which then upregulates the expression of many proteins required for cell proliferation, including cellular DNA replication proteins [11,29,76,100]. In addition, E7 may also play other roles in regulating the proliferation-associated cellular proteins. E7 has been shown to interact with the cellular cyclin E-cdk2 complex [60,118]. Whether this interaction has an effect on the phosphorylation of E1 by cyclin E-cdk2 is unknown. Obviously several viral proteins are required for various aspects of how viral DNA replication and protein expression are regulated in concert with the differentiation program of the host cells. The roles of the E6 and E7 proteins are the most direct examples of how the function of these other viral proteins, which may not be directly involved in the DNA replication machinery, can nonetheless play critical roles for PV DNA replication.

It is possible that HPV DNA replication in highly differentiated cells may occur differently than what has been seen in undifferentiated cell types. One recent study used the transient transfection PV DNA replication assay where the transfected cells were treated with agents to induce partially differentiated phenotypes. After partial differentiation, further PV DNA replication did not appear to be initiating specifically within the HPV origin region [30]. This could indicate that replication in these cells is no longer initiated from the origin, but randomly in the viral genome. An alternative hypothesis proposed by the authors of the study, is the intriguing possibility that as infected cells begin to differentiate, the mechanism of PV DNA replication switches to act through a modified rolling circle mechanism. This model would necessitate that PVs have a way of resolving long dsDNA PV genome concatamers into single dsDNA PV genome circles ready for viral packaging. Since host cells contain all the enzymatic machinery necessary for homologous recombination, it will be interesting to see whether PVs have evolved a way to recruit the cellular DNA recombination machinery to sites of PV DNA replication.

Therapeutic targets

A specific therapeutic agent for the treatment of HPV infection is not currently available. Traditional antiviral therapeutic targets are viral enzymes or virus-specific replication or transcription mechanisms. Since HPVs rely so heavily on cellular enzymes for these functions, few good therapeutic targets exist. The major HPV enzyme is E1. E1 ATPase and helicase assays have long been used to screen for antiviral drugs, and E1 is still a good potential therapeutic target [71]. Understanding the E1–E2 interaction and the interaction of these viral DNA replication proteins with host cell replication factors may provide further avenues for the design of therapeutic agents against HPV. The E2 peptide that inhibits the E1–E2 interaction and HPV DNA replication [43] demonstrates the viability of this approach.

Acknowledgements

This work was supported by grants GM56406 and AI01686 (to T.M.) from the National Institutes of Health.

References

1. Amin AA, Titolo S, Pelletier A, Fink D, Cordingley MG, Archambault J. Identification of domains of the HPV11 E1 protein required for DNA replication *in vitro*. Virology 2000; 272(1): 137–150.
2. Antson AA, Burns JE, Moroz OV, Scott DJ, Sanders CM, Bronstein IB, Dodson GG, Wilson KS, Maitland NJ. Structure of the intact transactivation domain of the human papillomavirus E2 protein. Nature 2000; 403(6771): 805–809.
3. Bellanger S, Demeret C, Goyat S, Thierry F. Stability of the human papillomavirus type 18 E2 protein is regulated by a proteasome degradation pathway through its amino-terminal transactivation domain. J Virol 2001; 75(16): 7244–7251.
4. Berg M, Stenlund A. Functional interactions between papillomavirus E1 and E2 proteins. J Virol 1997; 71(5): 3853–3863.
5. Bonne-Andrea C, Tillier F, McShan GD, Wilson VG, Clertant P. Bovine papillomavirus type 1 DNA replication: the transcriptional activator E2 acts *in vitro* as a specificity factor. J Virol 1997; 71(9): 6805–6815.
6. Bullock PA. The initiation of simian virus 40 DNA replication *in vitro*. Crit Rev Biochem Molec Biol 1997; 32(6): 503–568.
7. Chao SF, Rocque WJ, Daniel S, Czyzyk LE, Phelps WC, Alexander KA. Subunit affinities and stoichiometries of the human papillomavirus type 11 E1:E2:DNA complex. Biochemistry 1999; 38(14): 4586–4594.
8. Chen G, Stenlund A. Characterization of the DNA-binding domain of the bovine papillomavirus replication initiator E1. J Virol 1998; 72(4): 2567–2576.
9. Chen G, Stenlund A. The E1 initiator recognizes multiple overlapping sites in the papillomavirus origin of DNA replication. J Virol 2001; 75(1): 292–302.
10. Chen G, Stenlund A. Two patches of amino acids on the E2 DNA binding domain define the surface for interaction with E1. J Virol 2000; 74(3): 1506–1512.
11. Cheng S, Schmidt-Grimminger DC, Murant T, Broker TR, Chow LT. Differentiation-dependent up-regulation of the human papillomavirus E7 gene reactivates cellular DNA replication in suprabasal differentiated keratinocytes. Genes & Devel 1995; 9(19): 2335–2349.
12. Chiang CM, Dong G, Broker TR, Chow LT. Control of human papillomavirus type 11 origin of replication by the E2 family of transcription regulatory proteins. J Virol 1992; 66(9): 5224–5231.
13. Chow LT, Broker TR. *In vitro* experimental systems for HPV: epithelial raft cultures for investigations of viral reproduction and pathogenesis and for genetic analyses of viral proteins and regulatory sequences. Clinics Dermatol 1997; 15(2): 217–227.
14. Chow LT, Broker TR. Papillomavirus DNA replication. Intervirology 1994; 37(3–4): 150–158.
15. Clertant P, Seif I. A common function for polyoma virus large-T and papillomavirus E1 proteins? Nature 1984. 311(5983): 276–279.
16. Collins KL, Russo AA, Tseng BY, Kelly TJ. The role of the 70 kDa subunit of human DNA polymerase alpha in DNA replication. EMBO J 1993; 12(12): 4555–4566.
17. Conger KL, Liu JS, Kuo SR, Chow LT, Wang TS. Human papillomavirus DNA replication. Interactions between the viral E1 protein and two subunits of human dna polymerase alpha/primase. J Biol Chem 1999; 274(5): 2696–2705.

18. Cooper CS, Upmeyer SN, Winokur PL. Identification of single amino acids in the human papillomavirus 11 E2 protein critical for the transactivation or replication functions. Virology 1998; 241(2): 312–322.

19. Cueille N, Nougarede R, Mechali F, Philippe M, Bonne-Andrea C. Functional interaction between the bovine papillomavirus virus type 1 replicative helicase E1 and cyclin E-Cdk2. J Virol 1998; 72(9): 7255–7262.

20. Demeret C, Yaniv M, Thierry F. The E2 transcriptional repressor can compensate for Sp1 activation of the human papillomavirus type 18 early promoter. J Virol 1994; 68(11): 7075–7082.

21. Dixon EP, Pahel GL, Rocque WJ, Barnes JA, Lobe DC, Hanlon MH, Alexander KA, Chao SF, Lindley K, Phelps WC. The E1 helicase of human papillomavirus type 11 binds to the origin of replication with low sequence specificity. Virology 2000; 270(2): 345–357.

22. Dong G, Broker TR, Chow LT. Human papillomavirus type 11 E2 proteins repress the homologous E6 promoter by interfering with the binding of host transcription factors to adjacent elements. J Virol 1994; 68(2): 1115–1127.

23. Dornreiter I, Copeland WC, Wang TS. Initiation of simian virus 40 DNA replication requires the interaction of a specific domain of human DNA polymerase alpha with large T antigen. Molec Cell Biol 1993; 13(2): 809–820.

24. Dornreiter I, Erdile LF, Gilbert IU, von Winkler D, Kelly TJ, Fanning E. Interaction of DNA polymerase alpha-primase with cellular replication protein A and SV40 T antigen. EMBO J 1992; 11(2): 769–776.

25. Dornreiter I, Hoss A, Arthur AK, Fanning E. SV40 T antigen binds directly to the large subunit of purified DNA polymerase alpha. EMBO J 1990; 9(10): 3329–3336.

26. Enemark EJ, Chen G, Vaughn DE, Stenlund A, Joshua-Tor L. Crystal structure of the DNA binding domain of the replication initiation protein E1 from papillomavirus. Molec Cell 2000; 6(1): 149–158.

27. Fanning E, Knippers R. Structure and function of simian virus 40 large tumor antigen. Annu Rev Biochem 1992; 61: 55–85.

28. Ferguson MK, Botchan MR. Genetic analysis of the activation domain of bovine papillomavirus protein E2: its role in transcription and replication. J Virol 1996; 70(7): 4193–4199.

29. Flores ER, Allen-Hoffmann BL, Lee D, Lambert PF. The human papillomavirus type 16 E7 oncogene is required for the productive stage of the viral life cycle. J Virol 2000; 74(14): 6622–6631.

30. Flores ER, Lambert PF. Evidence for a switch in the mode of human papillomavirus type 16 DNA replication during the viral life cycle. J Virol 1997; 71(10): 7167–7179.

31. Fouts ET, Yu X, Egelman EH, Botchan MR. Biochemical and electron microscopic image analysis of the hexameric E1 helicase. J Biol Chem 1999; 274(7): 4447–4458.

32. Gilbert DM, Cohen SN. Bovine papilloma virus plasmids replicate randomly in mouse fibroblasts throughout S phase of the cell cycle. Cell 1987. 50(1): 59–68.

33. Gillette TG, Borowiec JA. Distinct roles of two binding sites for the bovine papillomavirus (BPV) E2 transactivator on BPV DNA replication. J Virol 1998; 72(7): 5735–5744.

34. Gillitzer E, Chen G, Stenlund A. Separate domains in E1 and E2 proteins serve architectural and productive roles for cooperative DNA binding. EMBO J 2000; 19(12): 3069–3079.

35. Gonzalez A, Bazaldua-Hernandez C, West M, Woytek K, Wilson VG. Identification of a

short, hydrophilic amino acid sequence critical for origin recognition by the bovine papillomavirus E1 protein. J Virol 2000; 74(1): 245–253.

36. Gopalakrishnan V, Khan SA. E1 protein of human papillomavirus type 1a is sufficient for initiation of viral DNA replication. Proc Nat Acad Sci USA 1994; 91(20): 9597–9901.

37. Gopalakrishnan V, Sheahan L, Khan SA. DNA replication specificity and functional E2 interaction of the E1 proteins of human papillomavirus types 1a and 18 are determined by their carboxyl-terminal halves. Virology 1999; 256(2): 330–333.

38. Gopalakrishnan V, Walker S, Khan SA. Stimulation of human papillomavirus type 1a DNA replication by a multimerized AT-rich palindromic sequence. Virology 1995; 214(1): 301–330.

39. Han Y, Loo YM, Militello KT, Melendy T. Interactions of the papovavirus DNA replication initiator proteins, bovine papillomavirus type 1 E1 and simian virus 40 large T antigen, with human replication protein A. J Virol 1999; 73(6): 4899–4907.

40. Holt SE, Schuller G, Wilson VG. DNA binding specificity of the bovine papillomavirus E1 protein is determined by sequences contained within an 18-base–pair inverted repeat element at the origin of replication. J Virol 1994; 68(2): 1094–1102.

41. Holt SE, Wilson VG. Mutational analysis of the 18-base–pair inverted repeat element at the bovine papillomavirus origin of replication: identification of critical sequences for E1 binding and *in vivo* replication. J Virol 1995; 69(10): 6525–6532.

42. Jian Y, Schmidt-Grimminger DC, Chien WM, Wu X, Broker TR, Chow LT. Post-transcriptional induction of p21cip1 protein by human papillomavirus E7 inhibits unscheduled DNA synthesis reactivated in differentiated keratinocytes. Oncogene 1998; 17(16): 2027–2038.

43. Kasukawa H, Howley PM, Benson JD. A fifteen-amino-acid peptide inhibits human papillomavirus E1E2 interaction and human papillomavirus DNA replication *in vitro*. J Virol 1998; 72(10): 8166–8173.

44. Kuo SR, Liu JS, Broker TR, Chow LT. Cell-free replication of the human papillomavirus DNA with homologous viral E1 and E2 proteins and human cell extracts. J Biol Chem 1994; 269(39): 24058–24065.

45. Lee D, Sohn H, Kalpana GV, Choe J. Interaction of E1 and hSNF5 proteins stimulates replication of human papillomavirus DNA. Nature 1999; 399(6735): 487–491.

46. Leng X, Ludes-Meyers JH, Wilson VG. Isolation of an amino-terminal region of bovine papillomavirus type 1 E1 protein that retains origin binding and E2 interaction capacity. J Virol 1997; 71(1): 848–852.

47. Lentz MR, Pak D, Mohr I, Botchan MR. The E1 replication protein of bovine papilloma-virus type 1 contains an extended nuclear localization signal that includes a p34cdc2 phosphorylation site. J Virol 1993. 67(3): 1414–1423.

48. Li R, Botchan MR. Acidic transcription factors alleviate nucleosome-mediated repres-sion of DNA replication of bovine papillomavirus type 1. Proc Nat Acad Sc USA 1994; 91(15): 7051–7055.

49. Li R, Botchan MR. The acidic transcriptional activation domains of VP16 and p53 bind the cellular replication protein A and stimulate *in vitro* BPV-1 DNA replication. Cell 1993; 73(6): 1207–1221.

50. Lin BY, Ma T, Liu JS, Kuo SR, Jin G, Broker TR, Harper JW, Chow LT. HeLa cells are phenotypically limiting in cyclin E/CDK2 for efficient human papillomavirus DNA replication. J Biol Chem 2000; 275(9): 6167–6174.

51. Liu JS, Kuo SR, Broker TR, Chow LT. The functions of human papillomavirus type 11 E1, E2, E2C proteins in cell-free DNA replication. J Biol Chem 1995; 270(45): 27283–27291.

52. Liu JS, Kuo SR, Makhov AM, Cyr DM, Griffith JD, Broker TR, Chow LT. Human Hsp70 and Hsp40 chaperone proteins facilitate human papillomavirus-11 E1 protein binding to the origin and stimulate cell-free DNA replication. J Biol Chem 1998; 273(46): 30704–30712.

53. Liu Z, Ghai J, Ostrow RS, Faras AJ. The expression levels of the human papillomavirus type 16 E7 correlate with its transforming potential. Virology 1995; 207(1): 260–270.

54. Lu JZ, Sun YN, Rose RC, Bonnez W, McCance DJ. Two E2 binding sites (E2BS) alone or one E2BS plus an A/T-rich region are minimal requirements for the replication of the human papillomavirus type 11 origin. J Virol 1993; 67(12): 7131–7139.

55. Lusky M, Hurwitz J, Seo YS. The bovine papillomavirus E2 protein modulates the assembly of but is not stably maintained in a replication-competent multimeric E1-replication origin complex. Proc Nat Acad Sci USA 1994; 91(19): 8895–8899.

56. Ma T, Zou N, Lin BY, Chow LT, Harper JW. Interaction between cyclin-dependent kinases and human papillomavirus replication—initiation protein E1 is required for efficient viral replication. Proc Nat Acad Sci USA 1999; 96(2): 382–387.

57. MacPherson P, Thorner L, Parker LM, Botchan M. The bovine papilloma virus E1 protein has ATPase activity essential to viral DNA replication and efficient transformation in cells. Virology 1994; 204(1): 403–408.

58. Mantovani F, Banks L. The interaction between p53 and papillomaviruses. Semin Cancer Biol 1999; 9(6): 387–395.

59. Masterson PJ, Stanley MA, Lewis AP, Romanos MA. A C-terminal helicase domain of the human papillomavirus E1 protein binds E2 and the DNA polymerase alpha-primase p68 subunit. J Virol 1998; 72(9): 7407–7419.

60. McIntyre MC, Ruesch MN, Laimins LA. Human papillomavirus E7 oncoproteins bind a single form of cyclin E in a complex with cdk2 and p107. Virology 1996; 215(1): 73–82.

61. McShan GD, Wilson VG. Casein kinase II phosphorylates bovine papillomavirus type 1 E1 in vitro at a conserved motif. J Gen Virol 1997; 78(Pt 1): 171–177.

62. McShan GD, Wilson VG. Contribution of bovine papillomavirus type 1 E1 protein residue 48 to replication function. J Gen Virol 2000; 81(Pt 8): 1995–2004.

63. Melendy T, Sedman J, Stenlund A. Cellular factors required for papillomavirus DNA replication. J Virol 1995; 69(12): 7857–7867.

64. Muller F, Sapp M. Domains of the E1 protein of human papillomavirus type 33 involved in binding to the E2 protein. Virology 1996; 219(1): 247–256.

65. Muller F, Seo YS, Hurwitz J. Replication of bovine papillomavirus type 1 origin-containing DNA in crude extracts and with purified proteins. J Biol Chem 1994; 269(25): 17086–17094.

66. Newhouse CD, Silverstein SJ. Orientation of a novel DNA binding site affects human papillomavirus-mediated transcription and replication. J Virol 2001; 75(4): 1722–1735.

67. Ozbun MA, Meyers C. Human papillomavirus type 31b E1 and E2 transcript expression correlates with vegetative viral genome amplification. Virology 1998; 248(2): 218–230.

68. Parenteau NL, Nolte CM, Bilbo P, Rosenberg M, Wilkins LM, Johnson WE, Watson S, Mason VS, Bell E. Epidermis generated in vitro: practical considerations and applications. J Cell Biochem 1991; 45(3): 245–251.

69. Park P, Copeland W, Yang L, Wang T, Botchan MR, Mohr IJ. The cellular DNA polymerase alpha-primase is required for papillomavirus DNA replication and associates with the viral E1 helicase. Proc Nat Acad Sci USA 1994; 91(18): 8700–8704.

70. Penrose KJ, McBride AA. Proteasome-mediated degradation of the papillomavirus E2-TA protein is regulated by phosphorylation and can modulate viral genome copy number. J Virol 2000; 74(13): 6031–6038.

71. Phelps WC, Barnes JA, Lobe DC. Molecular targets for human papillomaviruses: prospects for antiviral therapy. Antiviral Chem Chemother 1998; 9(5): 359–377.

72. Prives C. The replication functions of SV40 T antigen are regulated by phosphorylation. [Review]. Cell 1990. 61(5): 735–738.

73. Rangasamy D, Wilson VG. Bovine papillomavirus E1 protein is sumoylated by the host cell Ubc9 protein. J Biol Chem 2000; 275(39): 30487–30495.

74. Rangasamy D, Woytek K, Khan SA, Wilson VG. SUMO-1 modification of bovine papillomavirus E1 protein is required for intranuclear accumulation. J Biol Chem 2000; 275(48): 37999–38004.

75. Rocque WJ, Porter DJ, Barnes JA, Dixon EP, Lobe DC, Su JL, Willard DH, Gaillard R, Condreay JP, Clay WC, Hoffman CR, Overton LK, Pahel G, Kost TA, Phelps WC. Replication-associated activities of purified human papillomavirus type 11 E1 helicase. Prot Express Purific 2000; 18(2): 148–159.

76. Ruesch MN, Laimins LA. Initiation of DNA synthesis by human papillomavirus E7 oncoproteins is resistant to p21-mediated inhibition of cyclin E-cdk2 activity. J Virol 1997; 71(7): 5570–5578.

77. Russell J, Botchan MR. cis-Acting components of human papillomavirus (HPV) DNA replication: linker substitution analysis of the HPV type 11 origin. J Virol 1995; 69(2): 651–660.

78. Sakai H, Yasugi T, Benson JD, Dowhanick JJ, Howley PM. Targeted mutagenesis of the human papillomavirus type 16 E2 transactivation domain reveals separable transcriptional activation and DNA replication functions. J Virol 1996; 70(3): 1602–1611.

79. Sanders CM, Maitland NJ. Kinetic and equilibrium binding studies of the human papillomavirus type-16 transcription regulatory protein E2 interacting with core enhancer elements. Nucl Acids Res 1994; 22(23): 4890–4897.

80. Sanders CM, Stenlund A. Recruitment and loading of the E1 initiator protein: an ATP-dependent process catalysed by a transcription factor. EMBO J 1998; 17(23): 7044–7055.

81. Sanders CM, Stenlund A. Transcription factor-dependent loading of the E1 initiator reveals modular assembly of the papillomavirus origin melting complex. J Biol Chem 2000; 275(5): 3522–3534.

82. Sarafi TR, McBride AA. Domains of the BPV-1 E1 replication protein required for origin-specific DNA binding and interaction with the E2 transactivator. Virology 1995; 211(2): 385–396.

83. Sedman J, Stenlund A. Co-operative interaction between the initiator E1 and the transcriptional activator E2 is required for replicator specific DNA replication of bovine papillomavirus in vivo and in vitro. EMBO J 1995; 14(24): 6218–6228.

84. Sedman J, Stenlund A. The initiator protein E1 binds to the bovine papillomavirus origin of replication as a trimeric ring-like structure. EMBO J 1996; 15(18): 5085–5092.

85. Sedman J, Stenlund A. The papillomavirus E1 protein forms a DNA-dependent hexa-

meric complex with ATPase and DNA helicase activities. J Virol 1998; 72(8): 6893–6897.

86. Sedman T, Sedman J, Stenlund A. Binding of the E1 and E2 proteins to the origin of replication of bovine papillomavirus. J Virol 1997; 71(4): 2887–2896.

87. Seo YS, Muller F, Lusky M, Hurwitz J. Bovine papilloma virus (BPV)-encoded E1 protein contains multiple activities required for BPV DNA replication. Proc Nat Acad Sci USA 1993; 90(2): 702–706.

88. Smale ST, Tjian R. Inhibition of simian virus 40 DNA replication by specific modification of T-antigen with oxidized ATP. J Biol Chem 1986. 261(31): 14369–14372.

89. Spalholz BA, McBride AA, Sarafi T, Quintero J. Binding of bovine papillomavirus E1 to the origin is not sufficient for DNA replication. Virology 1993; 193(1): 201–212.

90. Steger G Ham J, Yaniv M. E2 proteins: modulators of papillomavirus transcription and replication. Meth Enzymol 1996; 274(2): 173–185.

91. Stenlund A. Papillomavirus DNA replication. Cold Spring Harbor Laboratory Press, Cold Spring Harbor, 1996.

92. Stubenrauch F, Colbert AM, Laimins LA. Transactivation by the E2 protein of oncogenic human papillomavirus type 31 is not essential for early and late viral functions. J Virol 1998. 72(10): 8115–81123.

93. Stubenrauch F, Lim HB, Laimins LA. Differential requirements for conserved E2 binding sites in the life cycle of oncogenic human papillomavirus type 31. 1998. J Virol 72(2): 1071–1077.

94. Sun Y, Han H, McCance DJ. Active domains of human papillomavirus type 11 E1 protein for origin replication. J Gen Virol 1998; 79(Pt 7): 1651–1658.

95. Sverdrup F, Khan SA. Two E2 binding sites alone are sufficient to function as the minimal origin of replication of human papillomavirus type 18 DNA. J Virol 69(2): 1995; 1319–1323.

96. Swindle CS, Engler JA. Association of the human papillomavirus type 11 E1 protein with histone H1. J Virol 1998; 72(3): 1994–2001.

97. Swindle CS, Zou N, Van Tine BA, Shaw GM, Engler JA, Chow LT. Human papillomavirus DNA replication compartments in a transient DNA replication system. J Virol 1999; 73(2): 1001–1009.

98. Syrjanen SM, Syrjanen KJ. New concepts on the role of human papillomavirus in cell cycle regulation. Ann Med 1999; 31(3): 175–187.

99. Thain A, Webster K, Emery D, Clarke AR, Gaston K. DNA binding and bending by the human papillomavirus type 16 E2 protein. Recognition of an extended binding site. J Biol Chem 1997; 272(13): 8236–8242.

100. Thomas JT, Hubert WG, Ruesch MN, Laimins LA. Human papillomavirus type 31 oncoproteins E6 and E7 are required for the maintenance of episomes during the viral life cycle in normal human keratinocytes. Proc Nat Acad Sci USA 1999; 96(15): 8449–8454.

101. Thomas M, Pim D, Banks L. The role of the E6–p53 interaction in the molecular pathogenesis of HPV. Oncogene 1999; 18(53): 7690–7700.

102. Titolo S, Pelletier A, Pulichino AM, Brault K, Wardrop E, White PW, Cordingley MG, Archambault J. Identification of domains of the human papillomavirus type 11 E1 helicase involved in oligomerization and binding to the viral origin. J Virol 2000; 74(16): 7349–7361.

103. Titolo S, Pelletier A, Sauve F, Brault K, Wardrop E, White PW, Amin A, Cordingley

MG, Archambault J. Role of the ATP-binding domain of the human papillomavirus type 11 E1 helicase in E2-dependent binding to the origin. J Virol 1999; 73(7): 5282–5293.

104. Ustav E, Ustav M, Szymanski P, Stenlund A. The bovine papillomavirus origin of replication requires a binding site for the E2 transcriptional activator. Proc Nat Acad Sci USA 1993; 90(3): 898–902.

105. Ustav M, Stenlund A. Transient replication of BPV-1 requires two viral polypeptides encoded by the E1 and E2 open reading frames. EMBO J 1991; 10(2): 449–457.

106. Ustav M, Ustav E, Szymanski P, Stenlund A. Identification of the origin of replication of bovine papillomavirus and characterization of the viral origin recognition factor E1. EMBO J 1991; 10(13): 4321–4329.

107. White PW, Pelletier A, Brault K, Titolo S, Welchner E, Thauvette L, Fazekas M, Cordingley MG, Archambault J. Characterization of recombinant HPV6 and 11 E1 helicases: effect of ATP on the interaction of E1 with E2 and mapping of a minimal helicase domain. J Biol Chem 2001; 276(25): 22426–22438.

108. Yang L, Li R, Mohr IJ, Clark R, Botchan MR. Activation of BPV-1 replication in vitro by the transcription factor E2. Nature 1991; 353(6345): 628–632.

109. Yang L, Mohr I, Fouts E, Lim DA, Nohaile M, Botchan M. The E1 protein of bovine papilloma virus 1 is an ATP-dependent DNA helicase. Proc Nat Acad Sci USA 1993; 90(11): 5086–5090.

110. Yang L, Mohr I, Li R, Nottoli T, Sun S, Botchan M. Transcription factor E2 regulates BPV-1 DNA replication in vitro by direct protein–protein interaction. Cold Spring Harbor Symp on Quantit Biol 1991; 56: 335–346.

111. Yasugi T, Benson JD, Sakai H, Vidal M, Howley PM. Mapping and characterization of the interaction domains of human papillomavirus type 16 E1 and E2 proteins. J Virol 1997; 71(2): 891–899.

112. Yasugi T, Howley PM. Identification of the structural and functional human homolog of the yeast ubiquitin conjugating enzyme UBC9. Nucl Acids Res 1996; 24(11): 2005–2010.

113. Yasugi T, Vidal M, Sakai H, Howley PM, Benson JD. Two classes of human papillomavirus type 16 E1 mutants suggest pleiotropic conformational constraints affecting E1 multimerization, E2 interaction, and interaction with cellular proteins. J Virol 1997; 71(8): 5942–5951.

114. Zou N, Lin BY, Duan F, Lee KY, Jin G, Guan R, Yao G, Lefkowitz EJ, Broker TR, Chow LT. The hinge of the human papillomavirus type 11 E2 protein contains major determinants for nuclear localization and nuclear matrix association. J Virol 2000; 74(8): 3761–3770.

115. Zou N, Liu JS, Kuo SR, Broker TR, Chow LT. The carboxyl-terminal region of the human papillomavirus type 16 E1 protein determines E2 protein specificity during DNA replication. J Virol 1998; 72(4): 3436–3441.

116. zur Hausen H. Immortalization of human cells and their malignant conversion by high risk human papillomavirus genotypes. Semin Cancer Biol 1999. 9(6): 405–411.

117. zur Hausen H. Papillomaviruses in human cancers. Proc Assoc Am Phys 1999; 111(6): 581–587.

118. Zwerschke W, Jansen-Durr P. Cell transformation by the E7 oncoprotein of human papillomavirus type 16: interactions with nuclear and cytoplasmic target proteins. Adv Cancer Res 2000; 78: 1–29.

Human Papillomaviruses
D.J. McCance (editor)

Human papillomavirus E6 protein interactions

Miranda Thomas*, David Pim and Lawrence Banks
International Centre for Genetic Engineering and Biotechnology, Padriciano 99, 34012 Trieste, Italy

Introduction

The evidence linking infection with Human Papillomaviruses (HPVs) to the development of a number of different human malignancies is now compelling. These include anogenital cancers, head and neck cancers and squamous cell carcinomas in renal transplant recipients (see Ref. [221] for review). However, the most numerically important malignant disease caused by HPV is carcinoma of the uterine cervix [220,222]. Therefore, most of the following discussion will focus on aspects of the role of HPV in the development of this tumour.

HPVs encode three oncoproteins, E5, E6 and E7 (see Ref. [149] for review). However, during the progression to malignancy HPV DNA sequences frequently become integrated into the host genome, and this results in large deletions of the viral genome [11,160]. As a result, the E5 gene is frequently lost, suggesting that its role in the malignant process is most likely during the early stages of cell immortalisation. In contrast, the E6 and E7 genes are invariably retained and expressed. Indeed, cell lines isolated from cervical cancers continue to express E6 and E7 many years after the initial immortalising events, indicating the continued importance of these proteins for maintenance of the transformed phenotype [4,15,161,168]. This hypothesis has now been widely confirmed by a number of studies which have shown that blocking E6 and E7 expression, by various means, results in these tumour-derived cell lines rapidly entering senescence and/or apoptosis. This has been achieved using either blocking peptides [28,136], antisense RNA [77,174,199,200] or ribozymes [2]. In all cases, inhibition of the expression of either, or both, of the two viral oncoproteins results in a cessation of transformed cell growth. This indicates a clear potential direction for effective chemotherapeutic intervention, and hence provides the impetus for understanding the mode of action of the viral oncoproteins.

Over 100 different HPV types have now been described, but only a small subset of these are associated with the development of human malignancies. One of the main challenges for many years has been to understand why only a small group of HPV types are associated with tumour development, whereas the vast majority of HPV types only induce benign proliferative lesions. Therefore during the following

---

*Corresponding author.

discussion we will attempt to highlight those aspects of E6 function which we believe are largely responsible for the differences in pathology observed between different HPV types.

HPV E6 and cell transformation.

Although the continued presence of E6 and E7 within cervical tumour-derived cell lines suggested a potential role for these viral proteins in the transformation process, the first direct evidence that they possessed intrinsic transforming activity came from studies in NIH3T3 cells [194,215]. In these studies it was found that the predominant transforming activity was encoded by the E7 gene [88,201], and that E6 was only weakly transforming in established rodent cells [20,162]. However, E6 from HPV-16 and HPV-18 can cooperate with an activated ras oncogene to transform primary rodent cells [118,146,178]. Interestingly, these studies provided the first evidence of clear biological differences in the modes of action of the viral oncoproteins derived from and high- and low-risk virus types. In these assays, high-risk HPV E6 and E7 proteins showed strong transforming activity, whereas the viral oncoproteins from low-risk HPVs were either non transforming or only weakly so [118,143,146,178,179].

Since PVs are species-specific, the most relevant transformation systems for assessing HPV oncoprotein function are those using human cells. It was shown that late-passage mammary epithelial cells could be immortalised by HPV E6 [13,203], but since the low-risk HPV E6 proteins are also active in this assay [14] it is not clear what transformation-relevant activities of E6 are being assayed in this system. The most meaningful *in vitro* assays of HPV transformation are obviously those using the target cells of the virus *in vivo*: primary human keratinocytes. A number of studies have shown that E6 and E7 from high-risk HPV types can cooperate to immortalise human foreskin, oral or cervical keratinocytes [45,91,159,163]. It is also clear that, as with primary rodent cells, the low-risk HPV E6 and E7 genes are inactive in such assays [142,211]. It should be emphasised that these immortalised keratinocytes are not fully transformed by E6 and E7, and cannot form tumours in nude mice. Only after extended passage of the cells in culture, or upon transfection of activated oncogenes, do the cells become tumorigenic [38,46,83], thereby reflecting the multi-step nature of cancer development *in vivo*.

Recently, additional insights into the activities of the two major HPV oncogenes were provided by an elegant series of studies in transgenic mice where E6 and E7 were expressed off the keratin 14 promoter. This promoter is activated in basal epithelium, and thus results in the E6 and E7 proteins being expressed in physiologic-ally relevant tissue [5,74,170]. Interestingly, these mice were found to develop epithelial neoplasias at a high frequency. However dissection of the individual roles of E6 and E7 in these transgenic mice proved to be most revealing. When E6 and E7 were expressed individually, each protein gave rise to skin tumours: when E7 was expressed alone tumours were induced with high frequency, they were, however, mainly benign differentiated tumours. In contrast, mice expressing only E6 in this system produced significantly fewer tumours, which were much more aggressive and prone to progression into metastatic cancer [171]. Further studies using chemical carcinogens indicate clearly that, while E7 is a promoter of immortalisation, E6 promotes progression into full malignancy [171]. Since it is the malignant aspect of cervical cancer that makes it fatal, it would seem that E6 might be the better target for chemotherapeutic intervention. Therefore in the following discussion we intend to review those activities of E6 that, in our view, are those that could be key inducers of transformation as well as contributing to the tendency of HPV-induced tumours to metastasise: these should represent the most effective targets for chemotherapeutic intervention.

Interaction of E6 with its cellular targets

The papillomavirus E6 proteins are approximately 150 amino acid polypeptides, having an apparent molecular weight of 18 kDa. They are characterised by the strict conservation of four metal binding Cys-x-x-Cys motifs, which permit the formation of two zinc fingers [17,31,68]. Mutants of E6 whose zinc fingers are disrupted are defective in almost all assays tried [89,165], indicating that this structural feature of E6 is vital for its correct action. It is also clear that high- and low-risk E6 proteins have much in common. Figure 1 shows a representation of the E6 sequence, high-lighting regions that are high-risk E6-specific, and those regions that are conserved amongst many different HPV E6 types. It is clear that the central portion of the molecule is highly conserved. Interestingly, however, the extreme amino- and carboxy-terminal portions of the protein are more divergent and, in particular, the E6 proteins of high-risk, mucosal-specific HPV types have an extended C-terminus containing a PDZ-binding motif, the implications of which will be discussed below.

Many different cellular targets of the high risk E6 proteins have now been described and it is impossible to effectively catalogue all of them, since many are the subjects of further investigation and their roles in the transformation process remain

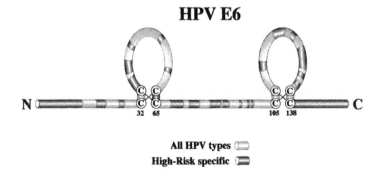

Fig. 1. Comparison of the HPV E6 amino acid sequences of high- and low-risk virus types. Regions in black show high-risk specific sequences. The remainder of the molecule shows a high degree of conservation across many different HPV types.

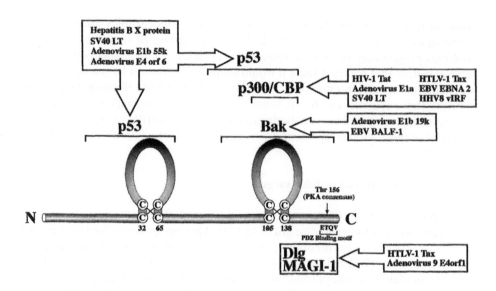

Fig. 2. A schematic diagram of the HPV-18 E6 protein and regions involved in interactions with some of its known cellular targets. Highlighted in boxes are those proteins also known to be targets of other viral transforming proteins, suggesting common pathways of viral transformation.

unknown. However, the analysis of potential common pathways of viral transformation can be useful, since many different human and animal viruses are now implicated in promoting cell transformation [see Ref. [221] for review]. When one analyses the cellular targets of other known viral oncoproteins, then a more limited subset of E6-interacting partners emerges. These are shown schematically in Fig. 2, together with their apparent sites of interaction on the HPV-18 E6 protein. Therefore we will largely focus our attention on this group of cellular targets that have currently known roles in regulating the processes that are required for the cell transformation induced by diverse viral families.

E6 and p53

The most famous, and most studied, target of high-risk HPV E6 is the tumour suppressor, p53 (see Ref. [187] for review). This protein was originally identified through its interaction with another viral oncoprotein, the SV40 large tumour antigen (TAg) [107,116], which binds to the p53 DNA binding domain, thus blocking p53's transactivation of target genes [154]. It was found to be similarly targeted by Adenovirus (Ad) E1B-55k protein [155], which binds the transactivation domain of p53 [115]. In addition, the Hepatitis B virus X protein prevents p53 transactivation by sequestering it in the cytoplasm [49]. These interactions result in the accumulation of p53 in stable and inactive complexes. When such complexes were sought in HPV-containing cells, little or no p53 protein was detected, despite there being plenty of p53 mRNA present [124,158]. It became clear from studies performed *in vitro* that E6 from HPV-16 and HPV-18 was able to bind to p53 and to induce its degradation via the ubiquitin pathway [157,205]. This activity of E6 explains why the levels of p53 in HPV-transformed cell lines are often so low [13]. It also explains the observations that p53 is wild type in such cell lines, and in HPV positive cervical tumours, while it is mutated in over 50% of all other tumours [37,158].

Interestingly, in addition to its ability to block p53 transactivation, Ad E1B-55k can associate with the Ad E4orf6 gene product to bring about p53 degradation [151,173], thereby highlighting an unexpected conservation of function between Adenoviruses and HPVs. E6 has also been shown in several studies to be able directly to inhibit p53 transactivation by inhibiting its DNA binding activity [109,146,186]. This is a function of E6 that is separate from its ability to degrade p53 and, intriguingly, low-risk E6 proteins have also been shown to be able to complex efficiently with p53 without targeting it for degradation [113]. Taken together, these data suggest that both high- and low-risk HPVs, as well as Adenoviruses, may actually require the presence of p53 at certain stages of the viral life-cycle. Since p53 has also been found in association with the replication complexes of a number of other viruses, including herpes simplex virus [208,217] and cytomegalovirus [55,132], it is possible that p53 is required in viral DNA replication. It has a $3'-5'$ exonuclease function, which could be used in a proofreading capacity [79] since Polymerase α, which is used in HPV replication [32,106,123], lacks this. Fuller support for this possibility comes from a recent report that has shown p53 to enhance the replicative fidelity of the HIV-1 reverse transcriptase [12]. In this context it is also interesting to note that in the case of HPVs, E2, the origin binding protein, has also been shown to interact with p53 [122].

The tumour suppressive activities of p53 are well documented (and well reviewed: see *Oncogene Reviews* issue no 18(53), 1999), and it is easy to see that the loss of p53 would result in an increased risk of transformation and malignant progression [42,93]. In addition, E6 has been shown to induce numerous chromosomal abnormalities [153,206] similar to those seen in cells lacking p53, suggesting that E6 inactivation of p53 may indeed contribute to genomic instability. One of the main activities of p53 is the induction of growth arrest and/or apoptosis [48,119,212] which of course are also non-permissive for viral replication. Papillomaviruses have very small genomes and do not express most of the proteins required for their replication; they are therefore dependent upon the replication machinery of the cell. However, their life-cycle requires that they replicate only in cells of the differentiating epithelium, which have exited the cell cycle and ceased DNA replication. The viral E7 protein acts in a number of ways to re-start the cell cycle; through release of E2F by its interaction with pRb; through blocking the inhibitory activities of p21 and p27; and by activating cyclin A and cyclin E promoters (see Chapter 5 and Refs. [133,149] for reviews). Reactivation of DNA replication by E7 in a differentiating cell activates p53 to initiate apoptosis, however, the ability of E6 to degrade p53 circumvents this potential problem for the virus.

It is interesting that E6 proteins of the low-risk HPV types either do not induce the degradation of p53, or do so at much lower efficiency than the high-risk E6 proteins [180]. The reason for this is probably the position of viral replication within the differentiating epithelium. The low-risk viruses initiate their replicative cycle perhaps more efficiently, but certainly lower in the stratified epithelium, than the high-risk viruses [43]. At this position the keratinocytes are probably still expressing some of the proteins required for viral DNA replication. In contrast the high-risk viruses do not initiate replication until later in the differentiation of the epithelium, when the cells have already switched off the expression of DNA replication proteins. Thus the high-risk virus E7 proteins have to exert stronger effects upon the cells to induce DNA replication [16,156]. This then causes, from the cells' points of view, inappropriate DNA replication; p53 is activated and E6 is then required to ablate this response.

For a long time it was widely assumed that the E6–p53 interaction was central to the ability of E6 to bring about cell transformation. However there is growing evidence that many other targets of E6 play a vital role in E6 induced transformation and malignancy [117,134,140,146]. Certainly, blocking the ability of E6 to target p53 for degradation is likely to have a negative impact on the growth of HPV transformed cells [21], although it is not clear how globally relevant this will be. Thus, in many cells derived from cervical tumours, the blocking of E6-induced degradation of p53 is not, alone, sufficient to induce high levels of p53 expression, indicating a lack of signalling to p53 in some cervical tumour-derived cell lines [120]. In addition, there are numerous reports of quite high levels of wild type p53 in cervical lesions at different stages of disease progression [33,114], suggesting that E6 is not degrading all p53 all of the time. This further questions the impact of an anti-E6–p53 therapeutic in terms of a natural viral infection.

Bak

Apoptosis, or programmed cell death, is a process that is hard-wired in the cell. It is necessary for the correct development and homeostasis of multicellular organisms [87,181], as well as being a defence against infection and neoplasia [192]. There are a number of different apoptosis-modulating pathways and the Bcl2 family proteins are critical death regulators immediately upstream of the mitochondria [1]. These proteins have been found in all multicellular eukaryotes tested and there are known to be at least 16 members in human cells [152,216]. This family of proteins have, variously, pro- or anti-apoptotic effects, and they can modulate these by forming hetero- or homo-dimers [67], which then form larger complexes. Monomeric inactive Bak resides in the mitochondrial membrane [209]; upon activation Bak forms active homo-dimers and higher order complexes [65,76,204] which then accelerate the opening of the voltage dependent anion channel (VDAC). This allows the rapid release of cytochrome C into the cytosol, which in turn leads to activation of the caspase cascade [166].

It has been shown that the Ad E1B-19k protein can in fact act as an anti-apoptotic member of the Bcl2 family (see Refs. [78,207] for review). Since there is often conservation of function between the DNA tumour virus proteins, it is perhaps not surprising that HPV E6 proteins target the pro-apoptotic Bak for degradation, *in vivo* [184,185]. Mutants of Bak that cannot be bound by E6 are resistant to degradation and, since they also have a considerably extended half-life even in the absence of HPV E6, it seems likely that E6 is enhancing the normal turnover of Bak in the cell. Interestingly, the equally pro-apoptotic Bax is not a target for degradation, although it has been mistakenly stated to be so. There are two possible reasons for this; the Bax gene is induced by p53 [129] and thus would not be likely to be a problem in E6-expressing cells. In addition, analyses of the tissue distribution of the two proteins have shown that Bak, but not Bax, is expressed at high levels in the upper epithelial layers, the site of HPV replication [101,102]. Interestingly more recent analyses have shown that the E6 proteins of cutaneous HPV can also target Bak for degradation [86], and that Bak levels do not rise in HPV-containing epithelium upon UVB irradiation. Thus Bak is a target of high and low-risk, mucosal and cutaneous, HPV types, indicating that this is a highly conserved function of the HPV E6 proteins.

Whether Bak represents a good target for anti-viral therapy remains to be seen. It might be the case for infections with cutaneous HPV types where many squamous cell carcinomas, unlike cervical carcinomas, have less than one copy of integrated HPV DNA sequence per cell [7,164]. Since this also implies that only fragments of HPV genome may be present, it indicates a very complex, and as yet insufficiently understood, pattern of events in the induction of these tumours. In cervical tumours the issue is not clear either, since Bak expression would normally be in the upper layers of the epithelium, at a stage of differentiation which is not reached during HPV-induced immortalisation and transformation. Therefore it seems more likely that Bak would represent a very good target for intervention at the stage of viral replication in normal infection, but is probably not so relevant in the later stages of the malignant disease.

Effects of E6 activity upon transcription—interaction with p300/CBP

The control of transcription from eukaryotic genes involves the assembly of multi-protein complexes, as well as RNA polymerase II. Transcriptional co-activators act as mediators between factors specific to a certain promoter and the general transcription machinery. The CBP/p300 family act in this way, and also have intrinsic histone acetylase activity, which is required for chromatin re-modelling [22,64]. They are thus extremely important in the control and correct functioning of the transcriptional machinery and it is perhaps not surprising that they are targeted by a number of tumour virus proteins (see Fig. 2). It has been shown that CBP/p300 targeting by SV40 TAg [47] and by Ad E1A protein [167,176] contributes to their transforming activities. Thus the reports that HPV-16 E6 and BPV-1 E6 proteins also interact with CBP/p300, and reduce its transcriptional activity [141,218,219] were extremely interesting, although there was little biological evidence to connect these observations with any effect upon transformation. However, it has recently been shown that both HPV-16 E6 and HPV-11 E6 can complement a mutant of Ad E1A, defective in p300 binding, in the transformation of primary epithelial cells [23]. Moreover this interaction appears to be independent of p53. Since the HPV-11, HPV-16 and BPV-1 E6 proteins have all been shown to interact with p300/CBP, it seems likely that this is an activity related to viral replication, perhaps altering promoter specificity to preferentially transcribe viral genes. This latter possibility seems particularly intriguing since the viral major transcriptional regulator, E2 has also recently been reported to function via interaction with p300 [110,121]. This raises the interesting possibility that E6 may interact with p300/CBP in order to down-regulate E2 transcriptional activity by a feedback mechanism.

Targets bound by E6 through helical domains

HPV E6 targets several cellular proteins for ubiquitin-mediated degradation by conjugating them with cellular ubiquitin ligases, the most well-known being E6-AP [81,82]. E6-AP is the prototype HECT domain protein and it is also known as UBE3A; mutations in this gene are responsible for Angelman syndrome, a serious developmental disease [96,125]. E6-AP was originally identified through its involvement in the E6-induced degradation of p53 [80,81]. In HPV-infected cells the complex between E6 and E6-AP forms an E3 ligase which recognises and ubiquitinates p53 [82]. In uninfected cells, p53 levels are normally also controlled through proteasome mediated degradation, using the ubiquitin-protein ligase MDM2 [75]. E6-AP is not thought to be involved in p53 degradation in the absence of E6, since blocking E6-AP activity with anti-sense oligonucleotides [21] or dominant-negative mutants [183] only results in increased p53 levels in HPV positive cells. In contrast, it appears that E6-AP does recognise Bak and c-Myc, and that E6 simply enhances the normal pathway by which these proteins' levels are controlled [69,184]. Interestingly E6 binding to E6-AP can also induce the self-ubiquitination of E6-AP itself [90], perhaps another example of feedback control.

HPV E6 binds to E6-AP through a linear helical motif on E6-AP, with a hydrophobic patch along one side that is essential for E6 binding [18,30,50]. This binding motif has been found on a number of other E6 binding partners, including E6BP/ERC55, MCM7 and paxillin [29,104,193], indicating that it is a common method by which E6 can interact with a number of cellular partners. However, the role of these binding partners in the oncogenic potential of HPV E6 is not yet clear.

The E6BP/ERC55 is a calcium-binding protein located in the endoplasmic reticulum and binding of E6 to this cellular protein is thought to explain the insensitivity to calcium-induced differentiation exhibited by E6-expressing cells [29,165]. Paxillin is involved in signal transduction between the actin cytoskeleton and focal adhesions and E6 binding results in the disruption of the actin cytoskeleton, though the mechanism by which this occurs is not entirely clear [193]. Although

E6BP/ERC55 and paxillin proteins have been reported to be bound by E6 proteins from high-risk (but not low-risk) HPVs, and from wild-type (but not transformation-defective) BPVs, neither protein is degraded in the presence of E6. Indeed, the binding of BPV E6 to paxillin competes with its binding to E6-AP [197]. Thus it seems likely that these interactions may be more relevant to BPV, which expresses considerably more E6 than the HPVs.

A similar region of E6 would also appear to be required for binding to MCM7. During the G1 phase of the cell cycle a complex, known as the licensing complex, binds to the cellular origins of DNA replication. Initiation of any origin cannot occur in the absence of the complex, but DNA replication also displaces it. Thus, replication from each cellular origin can only be initiated once in any cell cycle (see Ref. [190], for review). The MCM7 protein is one of a family of proteins involved in forming the licensing complex [191] and it was identified in a yeast two-hybrid screen as being a target for both high and low-risk HPV E6 proteins [104]. Although the consequences of the E6/MCM7 interaction are not yet clear, it seems probable that it contributes substantially to the greatly increased rate of viral DNA replication seen in infected cells as they begin to differentiate. Since the E6 protein from low-risk viruses also target MCM7, this is obviously not primarily responsible for any oncogenic effects, but it is possible that it contributes to the chromosomal instability seen in E6-expressing cells [53,206] which, in the context of high-risk E6-induced p53 degradation, could result in a secondary oncogenic effect.

The helical motifs involved in all of these interactions with E6 fall into the large family of LXXLL binding motifs, which are widely used in nuclear protein interactions [18,73,126]. They are also similar to those involved in p53 interactions with TAFII31 and MDM2 [195], and thus chemotherapeutic attempts to interfere with these reactions could have widespread toxic effects.

Myc

The c-Myc protein has also been shown to be a target for E6 enhanced degradation [69]. c-Myc is a cellular oncogene with an established association with oncogenesis [27]. However, as with E7, its overexpression gives rise to apoptosis [6,51] and, in keratinocytes, differentiation requires the down-regulation of c-Myc expression [57,144]. Since viral replication requires keratinocyte differentiation, the degradation of c-Myc appears to be logical. However, caution is required since E6 expression has been reported to upregulate c-Myc expression, by a post-transcriptional mechanism, in epithelial cells [202].

Interestingly, c-Myc's transcriptional activation of target promoters is mediated by hetero-dimerisation with the Max protein [3,103] and when Max instead hetero-dimerises with Mad, the same promoters are repressed [8,9]. This is reminiscent of the hetero-dimerisation control system used by the Bcl2 family in apoptosis, described above, and it indicates how E6 activity, by slightly altering the balance of key equilibria might have quite fundamental effects upon the fate of the host cell.

Effects of E6 on telomerase activity

Telomeres are repetitive nucleotide sequences found on the ends of mammalian chromosomes. They form complexes with DNA binding proteins, and these protect the chromosome ends, preventing end-to-end associations and fusions between chromosomes that might cause chromosome breakage during segregation [26,34]. Telomeres are normally eroded with each round of replication, and at a certain length may contribute to the onset of senescence [72]. Telomerase, which extends telomeres, is normally activated only in embryonal cells and adult germ cells [66,138]. However it has been shown that telomerase is activated in cells expressing HPV-16 E6 [98], although no increase in telomere stability was observed [177]. Paradoxically, telomere elongation, but no telomerase activity, was detected in E7-expressing cells [177]. Studies on cells immortalised by E6 and E7 only detected telomerase activation post-crisis [35,53], suggesting that it is a late activity during cellular transformation and that other epigenetic events might be required. More recently, it has been reported that the presence of E6 induces the expression of the telomerase reverse transcriptase gene, hTERT [61,139,198]. However, the precise mechanism for this is, as yet, unclear, although it has been reported that prolonged E6 expression results in the loss of a portion of chromosome 6 thought to encode an hTERT repressor [175]. It has also been reported that the hTERT gene is responsive to activation by c-Myc, amongst other factors [139], and also that E6 induction of hTERT may involve E6-AP, but not activation of c-Myc [61]. As discussed above, E6/E6-AP has been reported to degrade c-Myc protein [69], but the c-Myc gene has also been found to be activated by HPV integration in some cervical cancers [36,108]. From the data currently available, it appears that E6 activation of hTERT is most likely a late-stage marker of transformation and may involve a number of other factors, although a recent study suggests that E6 may be able to directly activate the hTERT promoter in transient transfection experiments [139]. However hTERT activation by E6 per se would not appear to be relevant during viral replication, and it most likely represents a readout of another, as yet unidentified, function of E6. Chemotherapeutic intervention directed at this activity of E6 could risk being too little, far too late.

Targets of E6 containing PDZ domains

As can seen from Figs. 2 and 3, the extreme carboxy terminal four amino acids of E6 (XTQV) are absolutely specific to high-risk, mucosal HPV E6s. These four amino acids comprise a consensus PDZ-binding domain, and, of all the protein interactions of E6, this is the most strictly high-risk specific. All the other activities of E6 can be detected, to a lesser or greater extent, amongst both high and low-risk HPV E6 proteins.

PDZ domains are stretches of 80–90 amino acids [52,150] that are bound with high affinity by the sequence T/SXV [172], which is usually found at the extreme carboxy terminal of their interacting proteins. The membrane-associated guanylate

		CONSENSUS PDZ DOMAIN-BINDING SEQUENCE	$...X^T_S XV$

HIGH-RISK MUCOSAL HPVs	HPV-18	CCNRARQERLQRRRETQV
	HPV-16	CC.....RSSRTRRETQL
	HPV-45	CCDQARQERLRRRRETQV
	HPV-31	CW......R.RPRTETQV
	HPV-33	CW......R.SRRRETAL
	HPV-35	CW......K.PTRRETEV
LOW-RISK MUCOSAL HPVs	HPV-6	CWTTCMEDMLP.......
	HPV-11	CWTTCMEDLLP.......
HIGH-RISK CUTANEOUS HPV	HPV-8	CKHLYHDW..........
LOW-RISK CUTANEOUS HPV	HPV-1	CRLYAI............

Fig. 3. Alignment of the carboxy terminal amino acids of high-risk E6 proteins, with low-risk HPV E6 proteins for comparison. Note the complete absence of PDZ-binding motif in low-risk E6. In addition, also note that HPV types 18, 45, 31 and 35 have a perfect PDZ binding consensus sequence, whereas HPV types 16 and 33 do not.

kinase homologues (MAGUKs) are large proteins with multiple protein–protein interaction domains, including PDZ domains, which are thought to mediate the formation of multiple-protein complexes. MAGUKs are located on the cell membrane, often at regions of cell–cell contact, including tight junctions [84,92,128], adherens junctions [54,137] and in the apical junctions of *C. elegans* [99,127]. Because of their location, it is thought that they may be involved in both inter-cell and intra-cell signalling, particularly since some also have nuclear localisation signals [41]. As well as high-risk E6, PDZ-binding motifs have been found in a number of other viral oncogenes, including the HTLV-1 Tax protein and the Ad9 E4-ORF1 gene product [97,111].

The Ad9 E4-ORF1 protein had been shown to interact with the human homologue of the Drosophila discs large protein, Dlg [10], and to sequester it in the membranous fraction of the cell [111]. Dlg is a putative tumour suppressor, since its deletion in Drosophila results in aberrant cytoskeletal organisation and in disruption of polarity in columnar epithelium [63]. Its role in maintaining cell polarity has been separated, by mutational analysis, from its role in controlling cellular proliferation [210]. In complex with the adenomatous polyposis coli (APC) protein, human Dlg represses the G1 to S transition [85], and its overexpression results in downregulation of β-catenin, and hence cell growth, via an APC-independent pathway [71]. HTLV-1 Tax protein has also been shown to bind Dlg and to counteract these activities [182]. Unlike the Tax and E4-ORF1 proteins, E6 is expressed at very low levels and cannot sequester Dlg, however it can target Dlg for ubiquitin-mediated degradation [59]. Dlg possesses three PDZ domains and, interestingly, the ability of E6 to target Dlg for degradation requires PDZ domain 2, which is the same PDZ domain recognised by APC [60]. This raises the intriguing possibility that E6, by degrading Dlg, may in fact be interfering with the normal functioning of the APC pathway which, along with p53, is one of the commonest pathways to be perturbed during the development of colon cancer [95,131].

More recently, it has been shown that the high-risk HPV E6s also induce the degradation of other MAGUK proteins. hScrib, which is the human homologue of Drosophila Scribble and is required for basolateral receptor localisation in epithelial cells [24,94] is targeted for degradation by E6 plus E6-AP [135]. Studies in Drosophila have shown that Scribble and Dlg can complement each other [25] and, since they are both targeted by HPV E6 proteins, this implies that E6 is perturbing a complex pattern of events regulating cell polarity and attachment in epithelial cells. Indeed, cells expressing E6 were shown to form weaker cell contacts and grow in a more disorganised fashion, whereas cells expressing E6 mutants defective for PDZ protein interactions retained normal aspects of cell polarity [135].

MUPP1 is another PDZ containing protein that has been recognised as a target of Ad9 E4ORF1 and HPV E6 [112]. MUPP1 is a multiple PDZ-containing protein [196]—it has 13 PDZ domains—and is thought to be involved in the selective targeting and assembly of signalling complexes [19]. In polarised epithelial cells it is localised exclusively at tight junctions, interacting with the Claudin-1 and JAM-1 (junctional adhesion molecule) proteins at adjacent PDZ domains—PDZ10 and PDZ9, respectively—and is also thought to be involved in cell polarity signalling [70].

The MAGI family of proteins is the most recently identified group of PDZ domain-containing targets for Ad9 E4ORF1 and HPV E6 [62,148,188]. These are MAGUKs with Inverted domain structure, having their guanylate kinase homology domains at the amino terminal end of the polypeptide. In addition MAGI-1, MAGI-2 and MAGI-3 each have two WW domains and at least five PDZ domains [41,213,214] giving them the potential to interact with a large number of other proteins, including β-catenin [40], mNET-1 [39] and PTEN [213,214]. As with Dlg and MUPP1, MAGI-1 is aberrantly sequestered in the cytoplasm by Ad9 E4-ORF1, and is targeted by high-risk HPV E6 proteins for ubiquitin-mediated degradation [62]. MAGI-2 and MAGI-3 are also targets for E6 induced degradation [189] and it is possible that they too are targets for Ad9 E4ORF1.

Comparison of the E6 induced degradation of target proteins has shown that the MAGI proteins are even more susceptible to degradation than p53, and certainly more so than Dlg or MUPP1 [112,148,188]. Apart from obvious differences in kinetics of degradation, there is other evidence to suggest that E6 does not degrade these proteins through identical pathways, invoking the possible involvement of other, as yet unidentified, ubiquitin ligases. This argument is supported by studies using chimaeric E6 molecules, where a short region of the carboxy terminus of HPV-18 E6 was fused in frame to the carboxy terminus of HPV-11 E6. This conferred PDZ protein-binding activity on the low-risk E6 protein, and allowed it to target Dlg for degradation [148]. In contrast, in the same assay, MAGI-1 was completely resistant to the chimaeric E6 protein. However, since HPV-11 E6 only binds weakly, if at all, to E6AP [81] this suggests that the ability of the chimaeric protein to degrade Dlg is indeed E6AP independent.

has also been shown that HPV-18 E6 binds to Dlg and to MAGI-1 more
ently than HPV-16 E6, and induces their degradation with concomitantly
er efficiency [148,188]. This is caused by a single amino acid difference in the
-binding motif at the carboxy terminus of E6, where HPV-18 E6 has a perfect
, and that of HPV-16 E6 is sub-optimal (see Fig. 3 for a comparison of high-risk
DZ binding domains). This contrasts with the case of p53, where HPV-16 E6 is
nore efficient [80,158], however it correlates very well with studies indicating
HPV-18-containing cervical tumours are more prone than HPV-16-containing
urs to recurrence and to metastasis (reviewed in Ref. [56]). This is the first
ng that correlates a clinical phenotype with a defined biochemical activity of an
protein, and indicates that the E6-induced control of MAGUK levels within the
ould be fundamental in the progression to malignancy.
n important feature of the E6-PDZ domain interaction is that it is highly
ed. Many of E6's other protein interactions involve large regions of the E6
in (see Fig. 2) and make molecular modelling and the rational design of
tors extremely difficult. In contrast the E6-PDZ interaction is mediated by a
amino acid motif, which can be disabled for binding either by point mutations
], or by simple phosphorylation of the threonine residue within the motif by
in kinase A [105]. Given these observations, and the fact that the structures of a
er of PDZ domains have been solved [44,100,130], the design of molecules that
interfere with these interactions seems highly feasible.
t present it is still not clear which of the above interactions will ultimately
ge as being important for E6's ability to contribute to malignant transformation.
inly, mutations within E6's PDZ-binding domain, in the context of the whole
result in weaker immortalising activity in keratinocytes than wild type E6
ins (C. Meyers, personal communication). However, based on the normal
on of these E6-MAGUK targets, one would most likely expect a more pro-
readout in assays that assess the contribution of E6 to malignant progression.
wait further developments in this area with great interest.

ficity—the right protein at the right time

the PDZ domain-containing proteins appear to impinge upon such a large
er of important pathways, it seems probable that blocking the interactions
en PDZ domains and PDZ-binding motifs might have extremely toxic effects.
ver, it is clear that there are additional layers of specificity controlling the
DZ interaction. Analysis of 34 PDZ domains from six proteins, including a
E6 target protein, has shown that only five domains can be bound by HPV E6
2,188,189]. This indicates the importance of peripheral amino acid sequences,
n the PDZ domain and its ligand. Specific binding to PDZ domains has also
shown with β-catenin, which binds to PDZ5 of MAGI-1 [40], with the tumour
essor PTEN, which binds to PDZ2 of MAGI-2 and MAGI-3 [213, 214], and

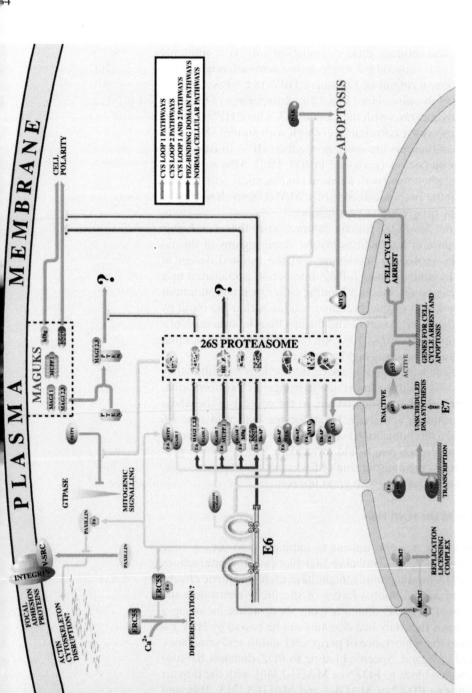

Opposite: Fig. 4. Cellular pathways perturbed by HPV E6 proteins during cell transformation.

domain interactions. Indeed, it has been shown that exogenously expressed MAGI-3 PDZ1 domain can block the *in vitro* and *in vivo* E6-induced degradation of Dlg, MAGI-2 and MAGI-3, thereby confirming proof-of-principle for this approach [189].

It is also becoming clearer that there are other factors that can determine the specificity of E6 interactions. We discussed earlier the possibility that a certain amount of p53 is required during the replication of viral DNA; obviously if that is the case there must be a mechanism for preventing certain E6 interactions. A potential means of doing this is via the E6* proteins; these are translated from alternatively spliced transcripts of the high-risk E6 gene [169]. They can bind to E6-AP or hetero-dimerise with E6, blocking their interaction [145,147] and thus preventing their inducing, or enhancing, the degradation of E6/E6-AP target proteins, of which p53 is the only one known to be resistant to E6-AP in the absence of E6.

Another important method of altering protein activity is by post-translational modifications such as phosphorylation, acetylation glycosylation, sumoylation and ubiquitination. Indeed as we have already seen, PKA phosphorylation of E6 has a profound effect upon its interaction with Dlg [105] and, most likely, with the other PDZ-containing targets. This implies that when E6 is phosphorylated by PKA it will not recognise its PDZ domain-containing substrates but will be available to carry out other activities, including p53 degradation [105]. In addition, recent studies have also shown that E6 is a substrate for phosphorylation by PKN [58]. At present this phosphorylation site on E6 is not known. However we await with great interest the elucidation of the consequences of the phosphorylation, with respect to the ability of E6 to recognise its target proteins.

Conclusions

When HPVs were first isolated and sequenced no one would have been able to predict the profound ways in which the HPV E6 proteins are capable of redirecting the cell's regulatory machinery in order to meet the requirements of viral replication. As can be seen from Fig. 4, the complexity of these interactions continues to develop, not only with the identification of new protein partners of E6, but also with the discovery of new cellular pathways and the elucidation of interconnections between apparently unconnected aspects of cellular homeostasis. Work with DNA tumour virus oncoproteins over the years has provided a wealth of information on central pathways regulating cell division. HPV E6 offers the exciting prospect of being able to utilise this information in the treatment of a major human cancer.

Acknowledgements

We are grateful to Craig Meyers, Oliviero Carugo and Ernesto Guccione for useful discussions. The authors gratefully acknowledge research support provided by the Associazione Italian per la Ricerca sul Cancro.

References

1. Adams JM, Cory S. The Bcl-2 protein family: arbiters of cell survival. Science 1998; 281: 1322–1326.
2. Alvarez-Salas LM, Cullinan AE, Siwkowski A, Hampel A, DiPaolo JA. Inhibition of HPV-16 E6/E7 immortalisation of normal keratinocytes by hairpin ribozymes. Proc Natl Acad Sci USA 1998; 95: 1189–1194.
3. Amati B, Dalton S, Brooks M, Littlewood T, Evan G, Land H. Transcriptional activation by the human c-Myc oncoprotein in yeast requires interaction with Max. Nature 1992; 359: 423–426.
4. Androphy EJ, Hubbert NL, Schiller JT, Lowy DR. Identification of the HPV-16 E6 protein from transformed mouse cells and human cervical carcinoma cell lines. EMBO J 1987; 6: 989–992.
5. Arbeit JM, Münger K, Howley P, Hanahan D. Progressive squamous epithelial neoplasia in K-14 human papillomavirus type–16 transgenic mice. J Virol 1994; 68: 4358–4368.
6. Askew D, Ashmun R, Simmons B, Cleveland J. Constitutive c-Myc expression in an IL-3-dependent myeloid cell line suppresses cell cycle arrest and accelerates apoptosis. Oncogene 1991. 6: 1915–1922.
7. Astori G, Lavergne D, Benton C, Hockmayr B, Egawa K, Garbe C, de Villiers EM. Human papillomaviruses are commonly found in normal skin of immunocompetent hosts. J Invest Dermatol 1998; 110: 752–755.
8. Ayer DE, Kretzner L, Eisenmann RN. Mad: a heterodimeric partner for Max that antagonises Myc transcriptional activity. Cell 1993; 72: 211–222.
9. Ayer DE, Lawrence QA, Eisenmann RN. Mad-Max transcriptional repression is mediated by ternary complex formation with mammalian homologs of yeast repressor, Sin3. Cell 1995; 80: 767–776.
10. Azim AC, Knoll JH, Marfatia SM, Peel DJ, Bryant PJ, Chishti AH. DLG1: a chromosome location of the closest human homologue of the Drosophila discs large tumour suppressor gene. Genomics 1995; 30: 613–616.
11. Baker CC, Phelps WC, Lindgren V, Braun MJ, Gonda MA, Howley PM. Structural and transcriptional analysis of human papillomavirus type 16 sequences in cervical carcinoma cell lines. J Virol 1987; 61: 962–971.
12. Bakhanashvili M. p53 enhances the replicative fidelity of DNA synthesis by human immunodeficiency virus type 1 reverse transcriptase. Oncogene 2001; 20: 7635–7644.
13. Band V, DeCaprio JA, Delmolino L, Kulesa V, Sager R. Loss of p53 protein in human papillomavirus type 16-immortalised human mammary epithelial cells. J Virol 1991; 65: 6671–6676.
14. Band V, Dalal S, Delmolino L, Androphy E. Enhanced degradation of p53 protein in HPV-6 and BPV-1 E6-immortalised human mammary epithelial cells. EMBO J 1993; 12: 1847–1852.
15. Banks L, Spence P, Androphy E, Hubbert N, Matlashewski G, Murray A, Crawford L. Identification of human papillomavirus type 18 E6 polypeptide in cells derived from human cervical carcinomas. J Gen Virol 1987; 68: 1351–1359.

16. Banks L, Edmonds C, Vousden KH. Ability of the HPV-16 E7 protein to bind RB and induce DNA synthesis is not sufficient for efficient transforming activity in NIH3T3 cells. Oncogene 1990; 5: 1383–1389.

17. Barbosa M, Lowy D, Schiller J. Papillomavirus polypeptides E6 and E7 are zinc binding proteins. J Virol 1989; 63: 1404–1407

18. Be X, Hong Y, Wei J, Androphy EJ, Chen JJ, Baleja JD. Solution structure determination and mutational analysis of the papillomavirus E6 interacting peptide of E6AP. Biochemistry 2001; 40: 1293–1299.

19. Becamel C, Figge A, Poliak S, Dumuis A, Peles E. Bockaert J, Lubbert H, Ullmer C. Interaction of serotonin 5-hydroxytryptamine type 2C receptors with PDZ10 of the multi-PDZ domain protein, MUPP1. J Biol Chem 2000; 276: 12974–12982.

20. Bedell MA, Jones KH, Grossman SR, Laimins LA. Identification of human papillomavirus type 18 transforming genes in immortalised and primary cells. J Virol 1989; 63: 1247–1255.

21. Beer-Romero P, Glass S, Rolfe M. Antisense targeting of E6AP elevates p53 in HPV-infected cells but not in normal cells. Oncogene 1997; 14: 595–602.

22. Berger SL. Gene activation by histone and factor acetyltransferases. Curr Opin Cell Biol 1999; 11: 336–341.

23. Bernat A, Massimi P, Banks L. Complementation of a p300/CBP defective binding mutant of Adenovirus E1a by the HPV E6 proteins. J Gen Virol 2002; 83: 829–833.

24. Bilder D, Perrimon N. Localisation of apical epithelial determinants by the basolateral PDZ protein Scribble. Nature 2000; 403: 676–680.

25. Bilder D, Li M, Perrimon N. Cooperative regulation of cell polarity and growth by *Drosophila* tumor suppressors. Science 2000; 289: 113–116.

26. Blackburn EH. Telomeres. Trends Biochem Sci 1991; 16: 378–381.

27. Bouchard C, Staller P, Eilers M. Control of cell proliferation by myc. Trends Cell Biol 1998; 8: 202–206.

28. Butz K, Denk C, Ullmann A, Scheffner M, Hoppe-Seyler F. Induction of apoptosis in human papillomavirus-positive cancer cells by peptide aptamers targeting the viral E6 oncoprotein. Proc Natl Acad Sci USA 2000; 97: 6693–6697.

29. Chen JJ, Reid C, Band V, Androphy E. Interaction of papillomavirus E6 oncoproteins with a putative calcium binding protein. Science 1995; 269: 529–531.

30. Chen JJ, Hong Y, Rustumzadeh E, Baleja JD, Androphy EJ. Identification of an alpha helical motif sufficient for association with papillomavirus E6. J Biol Chem 1998; 273: 13537–13544.

31. Cole ST, Danos O. Nucleotide sequence and comparative analysis of the human papillomavirus type 18 genome. Phylogeny of papillomaviruses and repeated structure of the E6 and E7 gene products. J Mol Biol 1987; 193: 599–608.

32. Conger KL, Lin JS, Kuo SR, Chow LT, Wang TS. Human papillomavirus DNA replication. Interactions between the viral E1 protein and two subunits of human DNA polymerase alpha/primase. J Biol Chem 1999; 274: 2696–2705.

33. Cooper K, Herrington CS, Evans MF, Gatter KC, McGee JO. p53 antigen in cervical condylomata, intraepithelial neoplasia, and carcinoma: relationship to HPV infection and integration. J Pathol 1993; 171: 27–34.

34. Coquelle A, Pipiras E, Toledo F, Buttin G, Debatisse M. Expression of fragile sites triggers intrachromosomal mammalian gene amplification and sets boundaries to early

amplicons. Cell 1997; 89: 215–225.

35. Coursen JD, Bennet WP, Gollahon L, Shay JW, Harris CC. Genomic instability and telomerase activity in human bronchial epithelial cells during immortalisation by human papillomavirus-16 E6 and E7. Exp. Cell Res 1997; 235: 245–253.

36. Couturier J, Sastre-Garau X, Schneider-Manoury S, Labib A, Orth G. Integration of papillomavirus DNA near myc genes in genital carcinomas and its consequence for proto-oncogene expression. J Virol 1991; 65: 4534–4538.

37. Crook T, Wrede D, Vousden K. p53 point mutation in HPV negative human cervical carcinoma cell lines. Oncogene 1991; 6: 873–875.

38. DiPaolo J, Woodworth C, Popescu MC, Notario V, Doniger J. Induction of human cervical squamous cell carcinoma by sequential transfection with human papillomavirus 16 DNA and viral Harvey ras. Oncogene 1989; 4: 395–399.

39. Dobrosotskaya IY. Identification of mNET1 as a candidate ligand for the first PDZ domain of MAGI-1. Biochem Biophys Res Commun 2001; 283: 969–975.

40. Dobrosotskaya IY, James GL. MAGI-1 interacts with beta-catenin and is associated with cell–cell adhesion structures. Biochem Biophys Res Commun 2000; 270: 903–909.

41. Dobrosotskaya I, Guy RK, James GL. MAGI-1, a membrane-associated guanylate kinase with a unique arrangement of protein–protein interaction domains. J Biol Chem 1997; 272: 31589–31597.

42. Donehower LA, Harvey M, Slagle BL, McArthur MJ, Montgomery Jr CA, Butel JS, Bradley A. Mice deficient for p53 are developmentally normal but susceptible to spontaneous tumours. Nature 1992; 356: 215–221.

43. Doorbar J, Foo C, Coleman N, Medcalf L, Hartley O, Prospero T, Napthine S, Sterling J, Winter G, Griffin H. Characterisation of events during the late stages of HPV16 infection *in vivo* using high affinity synthetic Fabs to E4. Virology 1997; 238: 40–52.

44. Doyle D, Lee A, Lewis J, Kim E, Sheng M, MacKinnon R. Crystal structures of a complexed and peptide-free membrane protein-binding domain: molecular basis of peptide recognition by PDZ. Cell 1996; 85: 1067–1076.

45. Dürst M, Dzarlieva-Petrusevka R, Boukamp P, Fusenig N, Gissmann L. Molecular and cytogenetic analysis of immortalised human primary keratinocytes obtained after transfection with human papillomavirus type 16 DNA. Oncogene 1987; 1: 251–256.

46. Dürst M, Gallahan D, Gilbert J, Rhim JS. Glucocorticoid enhanced neoplastic transformation of human keratinocytes by human papillomavirus type 16 and an activated ras oncogene. Virology 1989; 173: 767–771.

47. Eckner R, Ludlow JW, Lill NL, Oldread E, Arany Z, Modjtahedi N, DeCaprio JA, Livingston DM, Morgan JA. Association of p300 and CBP with simian virus 40 large T antigen. Mol Cell Biol 1996; 16: 3454–3464.

48. El Deiry W, Tokino T, Velculescu V, Levy D, Parsons R, Trent J, Lin D, Mercer W, Kinzler K, Vogelstein B. WAF-1, a potential mediator of p53 tumor suppression. Cell 1993; 75: 817–825.

49. Elmore LW, Hancock AR, Chang SF, Wang XW, Chang S, Callahan CP, Geller DA, Will H, Harris CC. Hepatitis B virus X protein and p53 tumor suppressor interactions in the modulation of apoptosis. Proc Natl Acad Sci USA 1997; 94: 14707–14712.

50. Elston R, Napthine S, Doorbar J. The identification of a conserved binding motif within human papillomavirus binding peptides. J Gen Virol 1998; 79: 371–374.

51. Evan G, Wyllie A, Gilbert C, Littlewood T, Land H, Brooks M, Waters C, Penn L,

Hancock D. Induction of apoptosis in fibroblasts by c-myc protein. Cell 1992; 69: 119–128

52. Fanning AS, Anderson JM. PDZ domains: fundamental building blocks in the organisation of protein complexes at the plasma membrane. J Clin Invest 1999; 103: 767–772.

53. Filatov L, Golubovskaya V, Hurt J, Byrd L, Phillips J, Kaufman W. Chromosomal instability, is correlated with telomerase erosion and inactivation of G2 checkpoint function in human fibroblasts expressing human papillomavirus type 16 E6 oncoprotein. Oncogene 1998; 16: 1825–1838.

54. Firestein BL, Rongo C. DLG-1 is a MAGUK similar to SAP97 and is required for adherens junction formation. Mol Biol Cell 2001; 12: 3465–3475.

55. Fortunato EA, Spector DH. p53 and RPA are sequestered in viral replication centers in the nuclei of cells infected with human cytomegalovirus. J Virol 1998; 72: 2033–2039.

56. Franco EL. Prognostic value of human papillomavirus in the survival of cervical cancer patients: an overview of the evidence. Cancer Epidemiol Biomarkers Prev 1992; 1: 499–504.

57. Freytag S, Dang C, Lee W. Definition of the activities and properties of c-Myc required to inhibit cell differentiation. Cell Growth Differ 1990; 1: 339–343.

58. Gao Q, Kumar A, Srinivasan S, Singh L, Mukai H, Ono Y, Wazer DE, Band V. PKN binds and phosphorylates human papillomavirus E6 oncoprotein. J Biol Chem 2000; 275: 14824–14830.

59. Gardiol D, Kühne C, Glaunsinger B, Lee SS, Javier R, Banks L. Oncogenic human papillomavirus E6 proteins target the discs large tumour suppressor for proteasome-mediated degradation. Oncogene 1999; 18: 5487–5496.

60. Gardiol D, Galizzi S, Banks L. Mutational analysis of the Discs Large tumour suppressor identifies domains responsible for HPV-18 E6 mediated degradation. J Gen Virol 2002; 83: 283–289.

61. Gewin L, Galloway DA. E-box dependent activation of telomerase by human papillomavirus type 16 E6 does not require induction of c-Myc. J Virol 2001; 75: 7198–7201.

62. Glaunsinger B, Lee SS, Thomas M, Banks L, Javier R. Interactions of the PDZ-protein MAGI-1 with adenovirus E4-ORF1 and high-risk papillomavirus E6 oncoproteins. Oncogene 2000; 19: 1093–1098.

63. Goode S, Perrimon N. Inhibition of patterned cell shape change and cell invasion by discs large during Drosophila oogenesis. Genes Dev 1997; 11: 2532–2544.

64. Goodman RH, Smolik S. CBP/p300 in cell growth, transformation and development. Genes Dev 2000; 14: 1553–1577.

65. Goping I, Gross A, Lavoie J, Nguyen M, Jemmerson R, Roth K, Korsmeyer S, Shore G. Regulated targeting of Bax to mitochondria. J. Cell Biol. 1998; 143: 207–215.

66. Greider CW, Blackburn EH. Identification of a specific telomere terminal transferase activity in Tetrahymena extracts. Cell 1985; 43: 405–413.

67. Gross A, McDonnell JM, Korsmeyer SJ. BCL-2 family members and the mitochondria in apoptosis. Genes Dev 1999; 13: 1899–1911.

68. Grossman SR, Laimins LA. E6 protein of human papillomavirus type 18 binds zinc. J Virol 1989; 65: 1247–1255.

69. Gross-Mesilaty S, Reinstein E, Bercovich B, Tobias K, Schwartz A, Kahana C, Ciechanover A. Basal and human papillomavirus E6 oncoprotein-induced degradation of Myc proteins by the ubiquitin pathway. Proc Natl Acad Sci USA 1998; 95: 8058–8063.

70. Hamazaki Y, Itoh M, Sasaki H, Furuse M, Tsukita S. Multi PDZ-containing protein 1

(MUPP1) is concentrated at tight junctions through its possible interaction with claudin-1 and junctional adhesion molecule (JAM). J Biol Chem 2001; 277: 455–461.

71. Hanada N, Makino K, Koga H, Morisaki T, Kuwahara H, Masuko N, Tabira Y, Hiraoka T, Kitamura N, Kikuchi A, Saya H. NE-dlg, a mammalian homolog of Drosophila dlg tumor suppressor, induces growth suppression and impairment of cell adhesion: possible involvement of down-regulation of beta-catenin by NE-dlg expression. Int J Cancer 2000; 86: 480–488.

72. Harley C, Futcher A, Greider CW. Telomeres shorten during ageing of human fibroblasts. Nature 1990; 345: 458–460.

73. Heery DM, Kalkhoven E, Hoarse S, Parker MG. A signature motif in transcriptional co-activators mediates binding to nuclear receptors. Nature 1997; 387: 733–736.

74. Herber R, Liem A, Pitot H, Lambert PF. Squamous epithelial hyperplasia and carcinoma in mice transgenic for the human papillomavirus type 16 E7 oncogene. J Virol 1996; 70: 1873–1881.

75. Honda R, Tanaka H, Yasuda H. Oncoprotein MDM2 is a ubiquitin ligase E3 for tumor suppressor p53. FEBS Lett 1997; 420: 25–27.

76. Hsu Y, Wolter K, Youle R. Cytosol-to-membrane redistribution of Bax and Bcl-XL during apoptosis. Proc Natl Acad Sci USA 1997; 94: 3668–3672.

77. Hu G, Liu W, Hanania EG, Fu S, Wang T, Deisseroth A. Suppression of tumorigenesis by transcription units expressing the antisense E6 and E7 messenger RNA (mRNA) for the transforming proteins of the human papillomavirus and the sense mRNA for the retinoblastoma gene in cervical carcinoma cells. Cancer Gene Ther 1995; 2: 19–32.

78. Huang DC, Cory S, Strasser A. Bcl-2, Bcl-XL and adenovirus protein E1B19kD are functionally equivalent in their ability to inhibit cell death. Oncogene 1997; 14: 405–414.

79. Huang P. Excision of mismatched nucleotides from DNA: a potential mechanism for enhancing DNA replication fidelity by the wild type p53 protein. Oncogene 1998; 17: 261–270.

80. Huibregtse J, Scheffner M, Howley P. A cellular protein mediates association of p53 with the E6 oncoprotein of human papillomavirus types 16 or 18. EMBO J 1991; 10: 4129–4135.

81. Huibregtse J, Scheffner M, Howley P. Cloning and expression of the cDNA for E6-AP, a protein that mediates the interaction of the human papillomavirus E6 oncoprotein with p53. Mol Cell Biol 1993; 13: 775–784.

82. Huibregtse J, Scheffner M, Howley P. Localisation of the E6-AP regions that direct human papillomavirus E6 binding, association with p53, and ubiquitination of associated proteins. Mol Cell Biol 1993; 13: 4918–4927.

83. Hurlin PJ, Kaur P, Smith PP, Peres-Reyes N, Blanton R, McDougall JK. Progression of human papillomavirus type 18-immortalised human keratinocytes to a malignant phenotype. Proc Natl Acad Sci USA 1991; 88: 570–574.

84. Ide N, Hata Y, Nishioka H, Hirao K, Yao I, Deguchi M, Mizoguchi A, Nishimori H, Tokino T, Nakamura Y, Takai Y. Localization of membrane-associated guanylate kinase (MAGI)-1/BAI-associated protein (BAP)1 at tight junctions of epithelial cells. Oncogene 1999; 18: 7810–7815.

85. Ishidate T, Matsumine A, Toyoshima K, Akiyama T. The APC/-hDLG complex negatively regulates cell cycle progression from the G0/G1 to S phase. Oncogene 2000; 19: 365–372.

86. Jackson S, Harwood C, Thomas M, Banks L, Storey A. Role of Bak in UV-induced apoptosis in skin cancer and abrogation by HPV E6 proteins. Genes Dev 2000; 14: 3065–3073.

87. Jacobson MD, Weil M, Raff MC. Programmed cell death in animal development. Cell 1997; 88: 347–354.

88. Kanda T, Furuno A, Yoshiike K. Human papillomavirus type 16 open reading frame E7 encodes a transforming gene for rat 3Y1 cells. J Virol 1988; 62: 610–613.

89. Kanda T, Watanabe S, Zanma S, Sato H, Furuno A, Yoshiike K. Human papillomavirus type 16 E6 proteins with glycine substituted for cysteine in metal binding motifs. Virology 1991; 185: 536–543.

90. Kao WH, Beaudenon SL, Talis AL, Huibregtse JM, Howley PM. Human papillomavirus type 16 E6 induces self-ubiquitination of the E6-AP ubiquitin-protein ligase. J Virol 2000; 74: 6408–6417.

91. Kaur P, McDougall JK. Characterisation of primary human keratinocytes transformed by human papillomavirus type 18. J Virol 1988; 62: 1917–1924.

92. Kawabe H, Nakanishi H, Asada M, Fukuhara A, Morimoto K, Takeuchi M, Takai Y. Pilt: a novel peripheral membrane protein at tight junctions in epithelial cells. J Biol Chem 2001; 276: 48350–48355.

93. Kemp CJ, Donehower LA, Bradley A, Balmain A. Reduction of p53 gene dosage does not increase initiation or promotion, but enhances malignant progression of chemically induced skin tumours. Cell 1993; 74: 813–822.

94. Kim SK. Polarized signaling: basolateral receptor localization in epithelial cells by the PDZ-containing proteins. Curr Opin Cell Biol 1997; 9: 853–859.

95. Kinzler KW, Vogelstein B. Lessons from hereditary colorectal cancer. Cell 1996; 87: 159–170.

96. Kishino T, Lalande M, Wagstaff J. UBE3A/E6-AP mutations cause Angelman syndrome. Nat Genet 1997; 15: 70–73.

97. Kiyono T, Hiraiwa A, Ishii S, Takahashi T, Ishibashi M. Binding of high-risk human papillomavirus E6 oncoproteins to a human homologue of the Drosophila discs large tumour suppressor protein. Proc Natl Acad Sci USA 1997; 94: 11612–11616.

98. Klingelhutz AJ, Foster SA, McDougall JK. Telomerase activation by the E6 gene product of human papillomavirus type 16. Nature 1996; 380: 79–82.

99. Köppen M, Simske JS, Sims PA, Firestein BL, Hall DH, Radice AD, Rongo C, Hardin JD. Cooperative regulation of AJM-1 controls junctional integrity in Caenorhabditis elegans epithelia. Nature Cell Biol 2001; 3: 983–991.

100. Kozlov G, Gehring K, Ekiel I. Solution structure of the PDZ2 domain from human phosphatase hPTP1E and its interactions with C-terminal peptides from the Fas receptor. Biochemistry 2000; 39: 2572–2580.

101. Krajewski S, Krajewska M, Shabaik A, Miyashita T, Wang HG, Reed JC. Immuno-histochemical determination of in vivo distribution of Bax, a dominant inhibitor of Bcl-2. Am J Path 1994; 145: 1323–1336.

102. Krajewski S, Krajewska M, Reed JC. Immunohistochemical analysis of in vivo patterns of Bak expression, a pro-apoptotic member of the Bcl-2 protein family. Cancer Res 1996; 56: 2849–2855.

103. Kretzner L, Blackwood EM, Eisenmann RN. Myc and Max proteins possess distinct transcriptional activities. Nature 1992; 359: 426–429.

104. Kühne C, Banks L. E3-ubiquitin ligase/E6-AP links multicopy maintenance protein 7 to the ubiquitination pathway by a novel motif, the L2G box. J Biol Chem 1998; 273: 34302–34309.

105. Kühne C, Gardiol D, Guarnaccia C, Amenitsch H, Banks L. Differential regulation of human papillomavirus E6 by protein kinase A: conditional degradation of human discs large protein by oncogenic E6. Oncogene 2000; 19: 5884-5891.

106. Kuo SR, Liu JS, Broker TR, Chow LT. Cell-free replication of the human papillomavirus DNA with homologous viral E1 and E2 proteins and human cell extracts. J Biol Chem 1994; 269: 24058–24065.

107. Lane DP, Crawford LV. T antigen is bound to a host protein in SV40 transformed cells. Nature 1979; 278: 261–263.

108. Lazo PA, DiPaolo JA, Popescu NC. Amplification of the integrated viral transforming genes of human papillomavirus 18 and its 5′-flanking cellular sequence located near the myc proto-oncogene in HeLa cells. Cancer Res 1989; 49: 4305–4310.

109. Lechner M, Laimins L. Inhibition of p53 DNA binding by human papillomavirus E6 proteins. J Virol 1994; 68: 4262–4275.

110. Lee D, Lee B, Kim J, Kim DW, Choe J. cAMP response element-binding protein binds to human papillomavirus E2 protein and activates E2-dependent transcription. J Biol Chem 2000; 275: 7045–7051.

111. Lee S, Weiss R, Javier R. Binding of human virus oncoproteins to hDlg/SAP97, a mammalian homologue of the Drosophila discs large tumour suppressor protein. Proc Natl Acad Sci USA 1997; 94: 6670–6675.

112. Lee S, Glaunsinger B, Mantovani F, Banks L, Javier R. The multi-PDZ domain protein MUPP1 is a cellular target for both human adenovirus E4-ORF1 and high-risk papillomavirus type 18 E6 oncoproteins. J Virol 2000; 74: 9680–9693.

113. Li X, Coffino P. High risk human papillomavirus E6 protein has two distinct binding sites within p53, of which only one determines degradation. J Virol 1996; 70: 4509–4516.

114. Lie AK, Skarsvag S, Skomedal H, Haugen OA, Holm R. Expression of p53, MDM2, and p21 proteins in high grade cervical intraepithelial neoplasia and relationship to human papillomavirus infection. Int J Gynecol Pathol 1999; 18: 5–11.

115. Lin J, Chen J, Elenbaas B, Levine AJ. Several hydrophobic amino acids in the p53 amino terminal domain are required for transcriptional activation, binding to mdm-2 and the adenovirus 5 E1B 55-kD protein. Genes Dev 1994; 8: 1235–1246.

116. Linzer DI, Levine AJ. Characterisation of a 54 kdalton cellular SV40 tumour antigen present in SV40-transformed cells and uninfected embryonal carcinoma cells. Cell 1979; 17: 43–52.

117. Liu Y, Chen JJ, Gao Q, Dalal S, Hong Y, Mansur CP, Band V, Androphy EJ. Multiple functions of human papillomavirus type 16 E6 contribute to the immortalisation of human mammary epithelial cells. J Virol 1999; 73: 7297–7307.

118. Liu Z, Ghai J, Ostrow RS, McGlennen RC, Faras AJ. The E6 gene of human papillomavirus type 16 is sufficient for transformation of baby rat kidney cells in cotransfection with activated Ha-ras. Virology 1994; 201: 388–396.

119. Lowe SW, Jacks T, Houseman DE, Ruley HE. Abrogation of oncogene associated apoptosis allows transformation of p53-deficient cells. Proc Natl Acad Sci USA 1994; 91: 2026–2030.

120. Mantovani F, Banks L. Inhibition of E6 induced degradation of p53 is not sufficient for stabilisation of p53 protein in cervical tumour derived cell lines. Oncogene 1999; 18: 3309–3315.

121. Marcello A, Massimi P, Banks L, Giacca M. Adeno-associated virus type 2 rep protein inhibits human papillomavirus type 16 E2 recruitment of the transcriptional co-activator p300. J Virol 2000; 74: 9090–9098.

122. Massimi P, Pim D, Bertoli C, Bouvard V, Banks L. Interaction between the HPV-16 E2 transcriptional activator and p53. Oncogene 1999; 18: 7748–7754.

123. Masterson PJ, Stanley MA, Lewis AP, Romanos MA. A C-terminal helicase domain of the human papillomavirus E1 protein binds E2 and the DNA polymerase alpha-primase p68 subunit. J Virol 1998; 72: 7407–7419.

124. Matlashewski G, Banks L, Pim D, Crawford L. Analysis of human p53 proteins and mRNA levels in normal and transformed cells. Eur J Biochem 1986; 154: 665–672.

125. Matsuura T, Sutcliffe J, Fang P, Galjaard R, Jiang Y, Benton C, Rommens J, Beaudet A. *De novo* truncating mutations in E6-AP ubiquitin-protein ligase (UBE3A) in Angelman syndrome. Nat Genet 1997; 15: 74–77.

126. McInerney EM, Rose DW, Flynn SE, Westin S, Mullen TM, Krones A, Inostroza J, Torchia J, Nolte RT, Assa-Munt N, Milburn MV, Glass CK, Rosenfeld MG. Determinants of coactivator LXXLL motif specificity in nuclear receptor transcriptional activation. Genes Dev 1998; 12: 3357–3368.

127. McMahon L, Legonis R, Vonesch JL, Labouesse M. Assembly of C elegans apical junctions involves positioning and compaction by LEET-413 and protein aggregation by the MAGUK protein DLG-1. J Cell Sci 2001; 114: 2265–2277.

128. Mino A, Ohtsuka T, Inoue E, Takai Y. Membrane-associated guanylate kinase with inverted orientation (MAGI)-1/brain angiogenesis inhibitor-1 associated protein (BAP1) as a scaffolding molecule for Rap small G protein GDP/GTP exchange protein at tight junctions. Genes Cells 2000; 5: 1009–1016.

129. Miyashita T, Reed JC. Tumor suppressor p53 is a direct transcriptional activator of the human bax gene. Cell 1995; 80: 293–299.

130. Morais Cabral JH, Petosa C, Sutcliffe MJ, Raza S, Byron O, Poy F, Marfatia SM, Chishti AH, Liddington RC. Crystal structure of a PDZ domain. Nature 1996; 382: 649–652.

131. Morin PJ, Vogelstein B, Kinzler KW. Apoptosis and APC in colorectal tumorigenesis. Proc Natl Acad Sci USA 1996; 93: 7950–7954.

132. Muganda P, Mendoza O, Hernandez J, Qian Q. Human cytomegalovirus elevates levels of the cellular protein p53 in infected cells. J Virol 1994; 68: 8028–8034.

133. Münger K, Basile JR, Duensing S, Eichten A, Gonzalez SL, Grace M, Zacny VL. Biological activities and molecular targets of the human papillomavirus E7 oncoprotein. Oncogene 2001; 20: 7888–7898.

134. Nakagawa S, Watanabe S, Yoshikawa H, Taketani Y, Yoshiike K, Kanda T. Mutational analysis of human papillomavirus type 16 E6 protein: transforming function for human cells and degradation of p53 *in vitro*. Virology 1995; 212: 535–542.

135. Nakagawa S, Huibregtse J. Human scribble (Vartul) is targeted for ubiquitin-mediated degradation by the high-risk papillomavirus E6 proteins and the E6-AP ubiquitin-protein ligase. Mol Cell Biol 2000; 20: 8244–8253.

136. Nauenburg S, Zwerschke W, Jansen-Dürr P. Induction of apoptosis in cervical carcinoma cells by peptide aptamers that bind to the HPV-16 E7 protein. FASEB J 2001; 15: 592–594.

137. Nishimura W, Iizuka T, Hirabayashi S, Tanaka N, Hata Y. Localisation of BAI-associated protein1/membrane associated guanylate kinase-1 at adherens junctions in normal rat kidney cells: polarised targeting mediated by the carboxy-terminal PDZ domains. J Cell Physiol 2000; 185: 358–365.

138. Noble JR, Rogan EM, Neumann AA, Maclean K, Bryan TM, Reddel RR. Association of extended *in vivo* proliferative potential with loss of p16INK4 expression. Oncogene 1996; 13: 1259–1268.

139. Oh ST, Kyo S, Laimins LA. Telomerase activation by human papillomavirus type 16 E6 protein: induction of human telomerase reverse transcriptase expression through Myc and GC-rich Sp1 binding sites. J Virol 2001; 75: 5559–5566.

140. Pan H, Griep AE. Temporally distinct patterns of p53-dependent and p-53-independent apoptosis during mouse lens development. Genes Dev 1995; 9: 2157–2169.

141. Patel D, Huang SM, Baglia LA, McCance DJ. The E6 protein of human papillomavirus type 16 binds to and inhibits co-activation by CBP and p300. EMBO J 1999; 18: 5061–5072.

142. Pecoraro G, Morgan D, Defendi V. Differential effects of human papillomavirus type 6, 16 and 18 DNAs on immortalisation and transformation of human cervical epithelial cells. Proc Natl Acad Sci USA 1989; 86: 563–567.

143. Phelps WC, Yee CL, Münger K, Howley PM. The human papillomavirus type 16 E7 gene encodes transactivation and transformation functions similar to those of adenovirus E1a. Cell 1988; 53: 539–547.

144. Pietenpol JA, Holt JT, Stein RW, Moses HL. Transforming growth factor 1 beta suppression of c-myc gene transcription: role in inhibition of keratinocyte proliferation. Proc Natl Acad Sci USA 1990; 87: 3758–3762.

145. Pim D, Banks L. HPV-18 E6*I protein modulates the E6-directed degradation of p53 by binding to full length HPV-18 E6. Oncogene 1999; 18: 7403–7408.

146. Pim D, Storey A, Thomas M, Massimi P, Banks L. Mutational analysis of HPV-18 E6 identifies domains required for p53 degradation *in vitro*, abolition of p53 transactivation *in vivo* and immortalisation of primary BMK cells. Oncogene 1994; 9: 1869–1876.

147. Pim D, Massimi P, Banks L. Alternatively spliced HPV-18 E6* protein inhibits E6-mediated degradation of p53 and suppresses transformed cell growth. Oncogene 1997; 15: 257–264.

148. Pim D, Thomas M, Javier R, Gardiol D, Banks L. HPV E6 targeted degradation of the discs large protein: evidence for the involvement of a novel ubiquitin-ligase. Oncogene 2000; 19: 719–725.

149. Pim D, Thomas M, Banks L. The function of the human papillomavirus oncogenes. In: RJA Grand (Ed), Viruses, Cell Transformation and Cancer. Elsevier, Amsterdam, 2001, pp. 145–192.

150. Ponting C, Phillips C. DHR domains in syntrophins, neural NO synthases and other intracellular proteins. Trends Biochem Sci 1995; 20: 102–103.

151. Querido E, Morrison MR, Chu-Phan-Dang H, Thirlwell SW, Boivin D, Branton PE, Morrison MR. Identification of three functions of Ad E4ORF6 protein that mediate p53 degradation by the E4ORF6-E1B55k complex. J Virol 2001; 75: 699–709.

152. Reed J. Bcl-2 family proteins. Oncogene 1998; 17: 3225–3236.

153. Reznikoff CA, Belair C, Savliev E, Zhai Y, Pfeifer K, Yeager T, Thompson KJ, DeVries S, Bindley C, Newton MA. et al. Long term stability and minimal genotypic and pheno-typic alterations in HPV16 E7-, but not E6-, immortalised human uroepithelial cells. Genes Dev 1994; 8: 2227–2240.

154. Ruppert JM, Stillman B. Analysis of a protein-binding domain of p53. Mol Cell Biol 1993; 13: 3811–3820.

155. Sarnow P, Ho YS, Williams J, Levine AJ. Adenovirus E1b-58kd tumor antigen and SV40 large tumor antigen are physically associated with the same 54kd cellular protein in transformed cells. Cell 1982. 28: 387–394.

156. Sato H, Furuno A, Yoshiike K. Expression of human papillomavirus type 16 E7 gene induces DNA synthesis of rat 3Y1 cells. Virology 1989; 168: 195–199.

157. Scheffner M, Werness B, Huibregtse J, Levine A, Howley P. The E6 oncoprotein encoded by human papillomavirus types 16 and 18 promotes the degradation of p53. Cell 1990; 63: 1129–1136.

158. Scheffner M, Münger K, Byrne JC, Howley PM. The state of the p53 and retinoblastoma genes in human cervical carcinoma cell lines. Proc Natl Acad Sci USA 1991; 88: 5523–5527.

159. Schlegel R, Phelps WC, Zhang YL, Barbosa M. Quantitative keratinocyte assay detects two biological activities of human papillomavirus DNA and identifies viral types associ-ated with cervical carcinoma. EMBO J 1988; 7: 3181–3187.

160. Schneider-Manoury S, Croissant O, Orth G. Integration of human papillomavirus type 16 DNA sequences: a possible early event in the progression of genital tumours. J Virol 1987; 61: 3295–3298.

161. Schwarz E, Freese U, Gissmann L, Mayer W, Roggenbuck B, Stremlau A, zur Hausen H. Structure and transcription of human papillomavirus sequences in cervical carcinoma cells. Nature 1985; 314: 111–114.

162. Sedman SA, Barbosa MS, Vass WC, Hubbert NL, Haas JA, Lowy DR, Schiller JT. The full length, E6 protein of human papillomavirus type 16 has transforming and trans-activating activities and cooperates with E7 to immortalise keratinocytes in culture. J Virol 1991; 65: 4860–4866.

163. Sexton CJ, Proby CM, Banks L, Stables JN, Powell K, Navsaria H, Leigh IM. Charac-terisation of factors involved in human papillomavirus type 16-mediated immortalisation of oral keratinocytes. J Gen Virol 1993; 74: 755–761.

164. Shamanin V, zur Hausen H, Lavergne D, Proby CM, Leigh IM, Neumann C, Hamm H, Goos M, Haustein UF, Jung EG et al. Human papillomavirus infections in non-melanoma skin cancers from renal transplant recipients and non-immunosuppressed patients. J Natl Cancer Inst 1996; 88: 802–811.

165. Sherman L, Schlegel R. Serum- and calcium-induced differentiation of human keratino-

cytes is inhibited by the E6 oncoprotein of human papillomavirus type 16. J Virol 1996; 70: 3269–3279.

166. Shimizu S, Narita M, Tsujimoto Y. Bcl-2 family proteins regulate the release of apoptogenic cytochrome C by the mitochondrial channel VDAC. Nature 1999; 399: 483–487.

167. Smith DH, Ziff E. The amino terminal region of adenovirus serotype 5 E1a protein performs two separate functions when expressed in primary baby rat kidney cells. Mol Cell Biol 1988; 8: 3882–3890.

168. Smotkin D, Wettstein F. Transcription of human papillomavirus type 16 early genes in a cervical cancer and a cancer-derived cell line, and identification of the E7 protein. Proc Natl Acad Sci USA 1986; 83: 4680–4684.

169. Smotkin D, Prokoph H, Wettstein FO. Oncogenic and non-oncogenic human genital papillomaviruses generate E7 mRNA by different mechanisms. J Virol 1989; 63: 1441–1447.

170. Song S, Pitot HC, Lambert PF. The human papillomavirus type 16 E6 gene alone is sufficient to induce carcinomas in transgenic animals. J Virol 1999; 73: 5887–5893.

171. Song S, Liem A, Miller JA, Lambert PF. Human papillomavirus type 16 E6 and E7 contribute differently to carcinogenesis. Virology 2000; 267: 141–150.

172. Songyang Z, Fanning AS, Fu C, Xu J, Marfatia SM, Chisti AH, Crompton A, Chan AC, Anderson JM, Cantley LC. Recognition of unique carboxyl-terminal motifs by distinct PDZ domains. Science 1997; 275: 73–77.

173. Steegenga WT, Riteco N, Jochemsen AG, Fallaux FJ, Bos JL. The large E1B protein together with the E4orf6 protein target p53 for active degradation in adenovirus infected cells. Oncogene 1998; 16: 349–357.

174. Steele C, Sacks PG, Adler-Storthz K, Shillitoe EJ. Effect on cancer cells of plasmids that express antisense RNA of human papillomavirus 18. Cancer Res 1992; 52: 4706–4711.

175. Steenbergen RD, Kramer D, Meijer CJ, Walboomers JM, Trott DA, Cuthbert AP, Newbold RF, Overkamp WJ, Zdzienicka MZ, Snijders PJ. Telomerase suppression by chromosome 6 in a human papillomavirus type 16-immortalised keratinocyte cell line and in a cervical cancer cell line. J Natl Cancer Inst 2001; 93: 865–872.

176. Stein RW, Corrigan M, Yaciuk P, Whelan J, Moran E. Analysis of E1A-mediated growth regulation functions: binding of the 300-kilodalton cellular product correlates with E1A enhancer repression function and DNA synthesis inducing activity. J Virol 1990; 64: 4421–4427.

177. Stöppler H, Hartmann DP, Sherman L. Schlegel, R. The human papillomavirus type 16 E6 and E7 oncoproteins dissociate cellular telomerase activity from the maintenance of telomere length. J Biol Chem 1997; 272: 13332–13337.

178. Storey A, Banks L. Human papillomavirus type 16 E6 gene cooperates with EJ-ras to immortalise primary mouse cells. Oncogene 1993; 8: 919–924.

179. Storey A, Pim D, Murray A, Osborn K, Banks L, Crawford L. Comparison of the in vitro transforming activities of human papillomavirus types. EMBO J 1988; 7: 1815–1820.

180. Storey A, Thomas M, Kalita A, Harwood C, Gardiol D, Mantovani F, Breuer J, Leigh I, Matlashewski G, Banks L. Role of a p53 polymorphism in the development of a human papillomavirus-associated cancer. Nature 1998; 393: 229–234.

181. Strasser A, Harris AW, Corcoran LM, Cory S. Bcl-2 expression promotes B- but not

T-lymphoid development in SCID mice. Nature 1994; 368: 457–460.

182. Suzuki T, Ohsugi Y, Uchida-Toita M, Akiyama T, Yoshida M. Tax oncoprotein of HTLV-1 binds to the human homologue of Drosophila discs large tumor suppressor protein, hDLG, and perturbs its function in cell growth control. Oncogene 1999; 18: 5967–5972.

183. Talis AL, Huibregtse JM, Howley PM. The role of E6AP in the regulation of p53 protein levels in human papillomavirus (HPV)-positive and HPV-negative cells. J Biol Chem 1998; 273: 6439–6445.

184. Thomas M, Banks L. Inhibition of Bak-induced apoptosis by HPV-18 E6. Oncogene 1998; 17: 2943–2954

185. Thomas M, Banks L. Human papillomavirus (HPV) E6 interactions with Bak are conserved amongst E6 proteins from high and low risk HPV types. J Gen Virol 1999; 80: 1513–1517.

186. Thomas M, Massimi P, Jenkins J, Banks L. HPV-18 E6 mediated inhibition of p53 DNA binding activity is independent of E6 induced degradation. Oncogene 1995; 10: 261–268.

187. Thomas M, Pim D, Banks L. Role of the E6-p53 interaction in the molecular pathogenesis of HPV. Oncogene 1999; 18: 7690–7700.

188. Thomas M, Glaunsinger B, Pim D, Javier R, Banks L. HPV E6 and MAGUK protein interactions: determination of the molecular basis for specific protein recognition and degradation. Oncogene 2001; 20: 5431–5439.

189. Thomas M, Laura R, Hepner K, Guccione E, Sawyers C, Lasky L, Banks L. Oncogenic human papillomavirus E6 proteins target the MAGI-2 and MAGI-3 proteins for degradation. Oncogene 2002; in press.

190. Thommes P, Blow JJ. The DNA replication licensing system. Cancer Surv 1997; 29: 75–90.

191. Thommes P, Kubota Y, Takisawa H, Blow JJ. The RLM-M component of the replication licensing system forms complexes containing all six MCM/P1 polypeptides. EMBO J 1997; 16: 3312–3319.

192. Thompson C. Apoptosis in the pathogenesis and treatment of disease. Science 1995; 267: 1456–1462.

193. Tong X, Howley PM. The bovine papillomavirus E6 oncoprotein interacts with paxillin and disrupts the actin cytoskeleton. Proc Natl Acad Sci USA 1997; 94: 4412–4417.

194. Tsunokawa Y, Takebe N, Kasamatsu T, Terada M, Sugimura T. Transforming activity of human papillomavirus type 16 DNA sequence in a cervical cancer. Proc Natl Acad Sci USA 1986; 83: 2200–2203.

195. Uesugi M, Verdine GL. The alpha helical FXXPhiPhi motif in p53: Taf interaction and discrimination by MDM2. Proc Natl Acad Sci USA 1999; 96: 14801–14806.

196. Ullmer C, Schmuck K, Figge A, Lubbert H. Cloning and characterisation of MUPP1, a novel PDZ domain protein. FEBS Lett 1998; 424: 63–68.

197. Vande Pol S, Brown MC, Turner CE. Association of Bovine Papillomavirus type 1 E6 oncoprotein with the focal adhesion protein paxillin through a conserved protein interaction motif. Oncogene 1998; 16: 43–52.

198. Veldman T, Horikawa I, Barrett JC, Schlegel R. Transcriptional activation of the telomerase hTERT gene by human papillomavirus type 16 E6 oncoprotein. J Virol 2001;

75: 4459–4466.

199. Venturini F, Braspenning J, Homann M, Gissmann L, Sczakiel G. Kinetic selection of HPV-16 E6 /E7-directed antisense nucleic acids: antiproliferative effects on HPV-16 transformed cells. Nucleic Acids Res 1999; 27: 1585–1592.

200. von Knebel Döberitz, M, Rittmuller C, zur Hausen H, Dürst M. Inhibition of tumorigenicity of cervical cancer cells in nude mice by HPV E6-E7 antisense RNA. Int J Cancer 1992; 51: 831–834.

201. Vousden KH, Doniger J, DiPaolo JA, Lowy DR. The E7 open reading frame of human papillomavirus type 16 encodes a transforming gene. Oncogene Res 1988; 3: 167–175.

202. Wang J, Xie LY, Allen S, Beach D, Hannon GJ. Myc activates telomerase. Genes Dev 1998; 12: 1769–1774.

203. Wazer DE, Liu XL, Chu Q, Gao Q, Band V. Immortalisation of distinct human mammary epithelial cell types by human papillomavirus 16 E6 or E7. Proc Natl Acad Sci USA 1995; 92: 3687–3691.

204. Wei MC, Lindsten T, Moothka VK, Weiler S, Gross A, Ashiya M, Thompson C, Korsmeyer SJ. tBID, a membrane-targeted death ligand, oligomerises BAK to release cytochrome c. Genes Dev 2000; 14: 2060–2071.

205. Werness B, Levine A, Howley P. Association of human papillomavirus types 16 and 18 E6 proteins with p53. Science 1990; 248: 76–79.

206. White AE, Livanos EM, Tlsty TD. Differential disruption of genomic integrity and cell cycle regulation in normal human fibroblasts by the HPV oncoproteins. Genes Dev 1994; 8: 666–677.

207. White, E. Regulation of the cell cycle and apoptosis by the oncogenes of adenovirus. Oncogene 2001; 20: 7836–7846.

208. Wilcock D, Lane DP. Localisation of p53, retinoblastoma and host replication protein at sites of viral replication in herpes-infected cells. Nature 1991; 349: 429–431.

209. Wolter KG, Hsu YT, Smith CL, Nechushtan A, Xi XG, Youle RJ. Movement of Bax from the cytosol to mitochondria during apoptosis. J Cell Biol 1997; 139: 1281–1292.

210. Woods DF, Wu JW, Bryant PJ. Localisation of proteins to the apico-lateral junctions of *Drosophila epithelia*. Dev Genet 1996; 20: 111–118.

211. Woodworth CD, Doniger J, DiPaolo JA. Immortalisation of human foreskin keratinocytes by various human papilloma virus DNAs corresponds to their association with cervical carcinoma. J Virol 1989; 63: 159–164

212. Wu X, Levine AJ. p53 and E2F-1 cooperate to mediate apoptosis. Proc Natl Acad Sci USA 1994; 91: 3602–3606.

213. Wu X, Hepner K, Castelino-Prabhu S, Do D, Kaye M, Yuan X-J, Wood J, Ross C, Sawyers CL, Whang YE. Evidence for regulation of the PTEN tumor suppressor by a membrane-localized multi-PDZ domain containing protein, MAGI-2. Proc Natl Acad Sci USA 2000; 97: 4233–4238.

214. Wu Y, Dowbenko D, Spencer S, Laura R, Lee J, Gu Q, Laskey L. Interaction of the tumor suppressor PTEN/MMAC with a PDZ domain of MAGI-3, a novel membrane associated guanylate kinase. J Biol Chem 2000; 275: 21477–21485.

215. Yasumoto S, Burkhardt AL, Doniger J, DiPaolo JA. Human papilloma virus type 16 DNA-induced malignant transformation of NIH3T3 cells. J Virol 1986; 57: 572–577.

216. Zamzami N, Brenner C, Marzo I, Susin SA, Kroemer G. Subcellular and submito-chondrial mode of action of Bcl2-like oncoproteins. Oncogene 1998; 16: 2265–2282

217. Zhong L, Hayward GS. Assembly of complete, functionally active herpes simplex virus DNA replication compartments and recruitment of associated viral and cellular proteins in transient cotransfection assays. J. Virol 1997; 71: 3146–3160.

218. Zimmermann H, Degenkolbe R, Bernard HU, O'Connor MJ. The human papilloma-virus type 16 E6 oncoprotein can downregulate p53 activity by targeting the trans-criptional coactivator CBP/p300. J Virol 1999; 73: 6209–6219

219. Zimmermann H, Koh CH, Degenkolbe R, O'Connor MJ, Muller A, Steger G, Chen JJ, Lui Y, Androphy EJ, Bernard HU. Interaction with CBP/p300 enables the bovine papillomavirus type 1 E6 oncoprotein to downregulate CBP/p300-mediated trans-activation by p53. J Gen Virol 2000; 81: 2617–2623.

220. zur Hausen H. Human papillomaviruses in the pathogenesis of anogenital cancer. Virology 1991; 184, 9–13.

221. zur Hausen H. Oncogenic DNA viruses. Oncogene 2001; 20: 7820–7823.

222. zur Hausen H, Schneider A. The role of papillomaviruses in human anogenital cancers. In: N Salzman and P Howley (Eds), The Papovaviridae, Vol. 2, Plenum Press, New York. 1987, pp. 245–263.

Human Papillomaviruses
D.J. McCance (editor)

The Biology of E7

Dennis J. McCance
Department of Microbiology & Immunology and the Cancer Center, University of Rochester, 601 Elmwood Avenue, Rochester, New York 14642, USA

Introduction

Papillomaviruses replicate in stratified epithelium and the major amplification step takes place in the upper part of the epithelium. This is the region of the epithelium where there is terminal differentiation and as such would not be conducive to viral DNA replication since components of the cellular replicative machinery would be low. In concert with other early region proteins like E6 and E5, the role of E7 is to stimulate cells which are programmed to differentiate, to re-enter the cell cycle and progress through to S-phase resulting in an adequate supply of the cell's replicative machinery for viral DNA replication. One of the consequences of this stimulation is that certain HPV types can immortalize primary cells in culture. As has been mentioned previously, high-risk HPVs can immortalize cultures of primary cells including the natural target cell of HPV, the keratinocyte. Inhibiting the activity of E6 or E7 by anti-sense, peptides or ribozymes will abrogate the immortalization phenotype of primary cultures, indicating that these proteins are necessary for the continuance of the immortalized state.

E7 is the major immortalizing protein of the high-risk types such as HPV-16 and 18, and can by itself immortalize primary human keratinocytes, although the process is much more efficient in the presence of E6. However, HPV-6, which produces benign lesions, must also stimulate cells into S-phase to replicate successfully, yet in culture the E7 protein from HPV-6 or even the full length genome is unable to immortalize keratinocytes and both fail to display other characteristics of oncogenic HPVs. One "black hole" of papilloma virology is how does HPV-6 achieve the same goals as HPV-16 of stimulating cells into S-phase, yet shows none of the tissue culture characteristics of the high-risk types.

HPV-16 E7 message is transcribed early in infection, producing low levels of protein which has been shown to interact with a number of cellular proteins, although the biological consequences of such interactions are not always clear. In this chapter I will initially discuss the interaction of E7 with various cellular proteins and then discuss the known biological consequences. Also I will concentrate on the activities of HPV-16 E7, which is typical of the high risk viruses, but where appropriate compare the functions of E7 from low risk HPV types, such as HPV-6.

Interaction of E7 with cellular proteins

Interaction of E7 with Rb

The major activity of E7 is its ability to modulate the transcription of various target genes in infected cells. The first cellular protein described to bind to E7 was retinoblastoma protein (Rb), a large cellular protein, which is involved in controlling the transcription of number of cellular genes, particularly those responsible for G1 to S-phase progression during the cell cycle [23,56]. There are three family members, Rb, p107 and p130, all of which can be bound by E7 [16,22]. For convenience E7 has been divided into three domains (Fig. 1), and the Rb binding domain is contained within the second region between amino acids 20–40 [7]. This region contains the typical LXCXE motif found in many proteins which bind Rb. Also contained in this region is a casein kinase consensus sequence [7], which can be phosphorylated *in vitro*, and although the biological activity of this phosphorylation has not been clearly defined, there are changes in the extent of phosphorylation of E7 during the cell cycle [51]. The N-terminal domain from amino acids 1–20 has not been extensively investigated and, at the time of writing, no cellular protein has been shown to bind to this region. However, deletion and point mutations within this region abrogate most of the biological properties of the protein, indicating an important function for this region [14,66], although all these mutants still bind the Rb family. Whether function(s) of the N-terminal domain depend(s) on binding a cellular protein, or is important for the structural integrity of E7, is at present unclear. Finally, there is the zinc finger domain in the C-terminal region and the structural integrity of this region is required for most, if not all, of the biological activity of E7 [3,15,25,32,37,54,66].

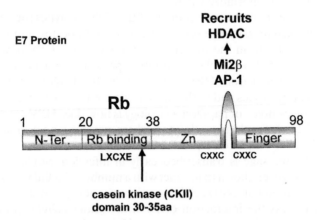

Fig. 1. Diagram of the three domains, CR1, CR2 and CR3 of HPV-16 E7 indicating the regions, which bind to cellular proteins. The CR1 domain is important for E7 function, although at present no cellular protein binds to this domain.

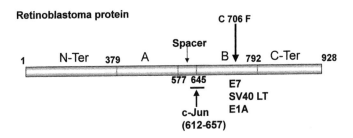

Fig. 2. Schematic of the retinoblastoma (Rb) protein and the region binding the various viral proteins, HPV-16 E7, SV40 large T (LT) and adenovirus E1a.

E7 binds to the B-pocket domain of Rb [34] through the LXCXE motif (Fig. 2). A mutation, C706P in the B-pocket inhibits binding of E7 and other viral proteins to Rb. Structural studies have shown that the bulky phenol group contained in the Phenylalanine disrupts the architecture of the pocket preventing binding of viral proteins [45]. As a tumor suppressor Rb inhibits the activity of a number of transcription factors, which are important for the transcription of proteins involved in G1 to S-phase progression and DNA synthesis. Normally Rb is phosphorylated late in G1, resulting in the abrogation of the repressive function and the activation of transcription factors such as the E2F family [21,46,47]. The binding of E7 to Rb appears to have the same effect on Rb as phosphorylation, as the repressive activity of Rb is abrogated [22, 34]. E7 from low-risk types such as HPV-6 can also bind to Rb although the level of binding is approximately 10 to 20% of that seen with the high-risk types [27]. It is unclear whether this level of binding has any biological relevance, however, experiments in tissue culture cells have shown that HPV-6 E7, unlike HPV-16 E7, does not activate E2F. The variance in binding efficiency can be accounted for by a single difference in the sequences next to their respective LXCXE motifs: HPV-6 E7 contains a glycine at amino acid 20, while HPV-16 E7 possesses an aspartic acid at the equivalent position [31,71]. Although the CR2 domain of HPV-16 E7 mediates high affinity binding to Rb, this region alone is unable to displace E2F from Rb and there appears to be a requirement for the CR3 domain. More recent studies have demonstrated that the CR3 region of E7 can also bind to Rb [65] and interacts with a region between 803 aa and 841 aa of Rb. Furthermore, *in vitro* and *in vivo* experiments indicate that CR3 is necessary for the derepression of Rb, leading to transcriptional activation. These results suggest that both the CR2 and CR3 domains are required for high affinity binding to Rb and the subsequent disruption of Rb-E2F complexes. While it would appear that the binding of Rb by E7 from high-risk types and the reduced binding of low risk types may account for the ability of HPV-16 to produce malignant disease, it should be pointed out that HPV-1, a virus infecting cutaneous surfaces, binds to Rb as well as HPV-16 E7, but this virus only produces benign disease and cannot immortalize primary cells [72]. These results indicate that other functions of E7 may be important for the phenotypes produced by high-risk viruses.

The mechanism by which Rb represses transcription has not been defined until recently. However, it has now been shown that the repressive activity is accomplished with the help of other cellular proteins, which are bound in a complex with Rb on the appropriate promoters. Two of these proteins are enzymes involved in chromatin remodeling. One is a family of at least 15 proteins called histone deacetylases (HDAC), which remove the acetyl group from the tails of histones resulting in condensed chromatin and inhibition of transcription [12,48]. The other family is a group called methyltransferases or methylases, which methylate histone tails also resulting in the repression of transcription. One family member, SUV39H1, has recently been shown to bind in the B-pocket of Rb [61]. It appears that deacetylation and methylation may be sequential events in the inhibition of transcription coordinated by Rb and it appears that Rb affects the status of chromatin as a means of inhibiting transcription.

Interaction of E7 with histone deacetylase activity

Chromatin remodeling, through various chemical alterations of histone proteins such as, acetylation, deacetylation, methylation and phosphorylation, is emerging as an important mechanism for the regulation of gene transcription. Actively transcribed genes show a high level of histone acetylation while repressed genes do not and are usually deacetylated and/or methylated [36]. Acetylation and deacetylation take place on histone tails at defined sites on lysines. Rb associates with a histone deacetylase (HDAC) activity and the complex represses transcription from Rb responsive promoters, such as E2F dependent genes. There are at least 15 HDAC proteins [49] divided into three classes and Rb has been shown to bind to at least two, HDAC-1 and 2. It appears that Rb does not bind directly to HDAC, but through an intermediary called the Rb-binding protein (RBP1 [44]), although the complex is formed through the B-pocket of Rb [12]. Since these Rb/HDAC containing complexes repress Rb dependent promoters, they play an important role in cell cycle regulation and are consequentially a target for viral oncogenes.

E7 has been shown to associate with deacetylase activity in the absence of Rb, although it does not directly bind HDAC proteins, but rather binds another protein called Mi2-beta, which binds to HDACs [13]. Mi2-beta is part of a large transcriptional repressive complex called NURD (Nucleosome Remodeling histone Deacetylatase complex [82]). Mutations in the CR3 domain (zinc finger), which do not affect Rb binding, or the integrity of the zinc finger structure, abrogate binding to Mi2-beta and thus to HDAC [13]. Therefore, using different domains, E7 binds to both Rb and Mi2-beta, which are members of independent chromatin remodeling complexes and the biological significance will be discussed later.

Interaction with AP-1 factors

Further attempts to identify Rb independent functions of E7 have led to the discovery that E7 can interact with members of the AP-1 family of transcription factors, including c-Jun, JunB, JunD and c-Fos [3]. AP-1 transcription factors appear to mediate early mitogenic effects and are further implicated in keratinocyte and myeloid cell differentiation [2,24]. Specific mutational analysis and binding data to c-Jun indicate that the E7 zinc finger motif, but not the Rb binding domain, is involved in this interaction. Moreover, E7 binds to 224-249 aa of c-Jun and can trans-activate transcription from a Jun responsive promoter. The E7/c-Jun inter-action was further demonstrated to be important in the ability of E7 to transform rat embryo fibroblasts (REF) in the presence of an activated *ras*.

It has also been observed that c-Jun binds to Rb, resulting in a Rb-dependent regulation of promoters containing AP-1 sites [59,62]. The interaction between Rb and c-Jun is mediated via the leucine zipper of c-Jun and the B pocket and c-terminus of Rb. c-Jun does not bind in the shallow groove of the Rb B-pocket, which binds E7, since the mutation C706F, which abrogates binding of E7 and other viral proteins, does not inhibit the binding of c-Jun [45]. c-Jun binds in a highly conserved domain which acts as an interface between the A and B-pockets of Rb [45] (Fig. 2). This area is also thought to be the binding site of the E2F family of transcription factors. Rb can recruit c-Jun to an AP-1 consensus site and activate transcription from c-Jun responsive promoters. Interestingly, it was demonstrated that the presence of E7 inhibited Rb activation of c-Jun transcription, and that this effect was mediated through the LXCXE motif of E7, suggesting that transcriptional down-regulation by E7 in this case is Rb dependent. However, since E7 and c-Jun bind to different regions of the B-pocket this inhibition cannot simply be due to steric hindrance. A similar situation is true for E7 and E2F in that the abrogation of E2F repression by Rb is more complex than just competition for the binding site on Rb. It was also shown that hypophosphorylated Rb is complexed to c-Jun in terminally differentiated keratinocytes but not in cycling cells. Finally, in terminally differentiated keratinocytes, the presence of E7 seems to cause a significant reduction in c-Jun levels.

The data discussed above suggest an intricate mode of action whereby HPV E7 can modulate both the process of cell cycle progression as well as cell differentiation, through interactions with AP-1 factors and Rb. On the one hand, by binding to c-Jun independently of Rb, E7 may potentiate the activation of genes involved in early cell cycle progression and thus promote S phase entry. On the other hand, by associating with both Rb and c-Jun, E7 may de-regulate keratinocyte differentiation via disruption of Rb/c-Jun complexes. As such, the differential targeting of AP-1 factors and Rb provides a potential mechanism used by E7 to uncouple differentiation from cell cycle progression.

Interaction with TBP and TAFs

The interactions described so far implicate the association of E7 with proteins that regulate the transcription of specific genes; namely those involved in cell cycling and differentiation. E7 however has been found to associate with members of the basal transcriptional machinery. E7 can bind to the TATA-binding protein (TBP [67]) and to TBP-associated factor-110 (TAF-110 [52]). Binding to TBP seems to require three domains of E7: the Rb binding domain, the CKII domain and the CR3 region [67]. The importance of these interactions is unclear, but they suggest that E7 may mediate its effects on gene transcription by targeting the more general transcription machinery.

Effects of E7 on cell growth

E7 has been demonstrated to alter growth phenotypes in a variety of cell types, the most biologically relevant of which is primary human foreskin keratinocytes (HFK). Immortalization of HFK correlates with oncogenicity of HPVs, with high-risk HPVs (e.g. HPV-16, 18, 31) able to efficiently induce immortalization and low-risk HPVs (e.g. HPV-6, 11) incapable of significantly extending HFK lifespan [8,35,53,68]. Expression of E7 from oncogenic HPV types is sufficient to immortalize HFK, though the immortalizing activity of E7 is more efficient in the context of E6 expression or the entire HPV genome [30]. Indeed, in the context of the full length genome, the presence of E7 is essential for HFK immortalization, as a premature stop codon mutation of the E7 open reading frame completely abrogates immortalization by full-length HPV16 genomes [37]. The CR1 and CR3 regions of E7 appear critical for its immortalizing activity, as mutations in these domains diminish the efficiency of E7-induced immortalization (Fig. 3) [37,77]. The immortalization ability of a selection of E7 mutants is shown in Table 1. Interestingly, Rb binding is not necessary for immortalization of HFK in the context of the whole genome as a mutation in amino acid 24 (C24G] in the LXCXE motif of HPV-16, which abrogates Rb binding, resulted in immortalization [37]. This suggests that interaction with Rb may be dispensable for immortalization in cell culture. However, it is not clear if Rb binding is required in the epithelium for viral DNA replication and production of premalignant lesions, although work with the cottontail rabbit papillomavirus (CRPV) showed that the full length genome, with the equivalent mutation in E7, was

E7 Protein	CR1	CR2	CR3
Rb binding/E2F activation		++++	+++
AP1 binding		+++	+++
HDAC association			+++
Transformation (rodent cells)	+++	++++	++
Immortalization	++++	+	+++

Fig. 3. Shows the biological functions of E7 and the domains, which are important for each function. ++++Means the region is absolutely required; +means the region is not required.

Table 1

Immortalization of human primary keratinocytes by HPV-16 E7 and mutants[a]

E7	Binding Rb	Binding HDAC activity	Degradation of Rb	Immortalization by E7 alone	Immortalization in co-operation with E6
Wild type	++++	++++	++++	++[b]	++++[b]
H2P	++++	++++	–[c,d]	–	–
C24G	–	++++	–[d]	–	++
L67R	++++	–	NT[e]	–	++++
S71C	++++	++++	NT[e]	++	++++
R77G	++++	++++	NT[e]	++	++++
LL82/83RR	++++	–	NT[e]	–	++++

[a]The number indicates the amino acid mutated.
[b]++++Immortalization in all cases; ++immortalization in approx. 50% of cases.
[c]Unable to immortalize.
[d]Taken from Helt, 2001 [23].
[e]Not tested.

able to produce wart lesions in domestic rabbits [17], although no information on whether these lesion progressed to malignancy was presented. It should be remembered that CRPV when injected into the skin of domestic rabbits results in warts with a high rate of progression to malignant squamous cell carcinomas. These results suggest that the binding of Rb may be dispensable for the production of premalignant disease, but leaves open the question whether binding is necessary for malignant conversion.

E7 can also induce cellular transformation in several assays. HPV-16 E7 can induce anchorage-independent growth in NIH3T3 and, in cooperation with an activated *ras*, induce focus formation in rat embryo fibroblasts (REF), baby rat kidney cells, and various rodent fibroblast cell lines [55]. Similar to HFK and REF immortalization, E7 transformation of rodent cells correlates with oncogenic potential of HPVs, and chimeric molecules (utilizing domains of E7 from oncogenic HPV-16 and benign HPV-6 in fusion constructs) indicate that the CR1 and CR2 regions of HPV-16 E7 are essential for this phenotype [57,77]. Mutational analysis has corroborated the importance of the CR1 and CR2 domains and shown that Rb binding is required [5,14,15,31,66]. In addition, mutation of the zinc-finger (CR3) reduces the transformation potential of E7; however, since the CR3 mutations disrupted the zinc finger structure in these studies, it is not clear whether these mutations disrupt CR3-specific activities necessary for transformation or simply alter the structure of E7 [15,25,54]. Therefore, it appears that overlapping but also separate domains of E7 are essential for rodent cell transformation and HFK immortalization and it is unclear which of these biological functions are important for malignant conversion by HPV-16.

In addition, to its immortalizing and transforming activities, E7 from HPV-16 but not from HPV-6, also abrogates several growth arrest signals including exposure to transforming growth factor-β, DNA damage, serum deprivation, anchorage-independent suspension, and suprabasal differentiation [5,18,73]. Although the precise mechanisms by which these signals lead to arrest are not completely defined, E7 has been shown to interact with several cellular factors which control transcription and cell cycle regulation and in the next two section, these interactions will be discussed.

Effects of E7 on cellular transcription

It is becoming increasingly clear that E7 mediates many of its effect on cell growth via the modulation of gene transcription and the cell phenotypes discussed in the last section are probably accounted for by regulation of transcription. E7 has been found to interact with several proteins believed to affect the transcription of genes involved in cell cycle progression and/or cellular differentiation. The significance of these interactions is discussed below.

As mentioned above, the most well characterized property of E7 is its ability to bind to the retinoblastoma tumor suppressor protein (Rb [23,56]). The fact that E7 shares this property with other viral oncoproteins, such as the adenovirus E1A and simian virus 40 large T antigen, suggests that tumor viruses possess evolutionary conserved attributes, and underscores the importance of Rb binding in the natural history of virus infection. The retinoblastoma protein family members play a central role in the regulation of the eukaryotic cell cycle. Specifically, in its hypophosphorylated state, Rb can bind to transcription factors such as the E2F family members, and repress the transcription of particular genes. As cells progress from G0 through G1 and into S phase, Rb family members become progressively hyperphosphorylated by G1 cyclin–cyclin dependent kinases, such as cyclin D1/cdk4/6 and cyclin E/cdk2. Phosphorylation results in the release of the transcription factor E2F, which in turn activates genes involved in DNA synthesis and cell cycle progression [21]. Since E7 is able to bind to hypophosphorylated Rb, it is believed that E7 can prematurely induce cells into S phase by disrupting Rb–E2F complexes [34,65,80]. Interaction with Rb is primarily mediated through amino acid sequences contained in the conserved amino terminal or CR2 region of E7, although the CR3 domain is required for derepression of promoters. The CR2 region of E7 binds to Rb and its family members, p107 and p130, through the sequence motif LXCXE [7,16,22]. The LXCXE motif of E7 has been shown to specifically bind to one Rb pocket region (B-pocket), between amino acids 649 and 772 [34]. Moreover, the LXCXE amino acid sequence is found within other viral oncoproteins such as E1A and SV40 LTAg, as well as many cellular Rb binding proteins including cyclin D1, cyclin D2, cyclin D3, BRG1 (a human homolog of SWI/SNF proteins in yeast), and histone deacetylases-1 and 2 [12,19,20,48]. The high level of conservation of this motif among viral and cellular proteins suggests that viral oncoproteins such as E7 may compete with cellular proteins for binding to Rb.

Recent developments in the Rb and E2F literature highlight the complexity and importance of different Rb family members. For example, p130/E2F4 complexes are the most predominant and are present in quiescent or differentiated cells, while p107/E2F1-3 and Rb/E2F1-3 can be found in cells entering G1 and S phase [21]. Moreover, p107 and p130 are required for the regulation of different subsets of genes. Considering the ability of E7 to interact with all known Rb family members and the latter's involvement in both differentiation and cell proliferation, it is tempting to suggest a paradigm whereby E7 can uncouple the process of differentiation from cell cycle progression by modulating the transcription of different subsets of genes. This in turn would establish an environment that is more conducive to viral replication.

How do the Rb family members repress transcription and how does E7 relieve that repression? Since Rb binds to the transcriptionally active domain of E2F, it was thought that Rb might repress transcription by inhibiting the ability of E2F to interact with the basal transcription machinery of the cell. Recently, however, Rb has been shown to interact with histone deacetylases [12,48] and a methylase (SUV39H1 [61]), enzymes that remodel chromatin and so affect transcription. Actively transcribed regions of DNA show a high level of histone acetylation and low methylation while repressed genes exhibit the opposite profile. In fact there is evidence accumulating which suggests there is a histone code, which determines the structure of chromatin and therefore whether a gene is transcribed or not [36]. Recent work has shown an interesting Rb-dependent sequence of events in this histone code hypothesis. This work has shown that for repression of transcription, lysine 9 on the tail of histone 3 is deacetylated and then methylation takes place on the same amino acid [61]. There is evidence that Rb/methylase complex cannot methylate histone 3 on lysine 9 when the lysine is already acetylated. Therefore, to repress an active promoter Rb/HDAC may deacetylate lysine 9 on the histone 3 tail and once deacetylated, Rb may then recruit the methylase to methylate lysine 9. Methylation of lysine 9 then specifically binds the Heterochromatin Protein (HP-1) protein, which is known to be involved in the condensation of sections of DNA (Fig. 4) [6]. Recent research suggests that histone methylases may also be involved in controlling DNA methylation, another negative signal for promoter regulation [78]. Therefore, there is a

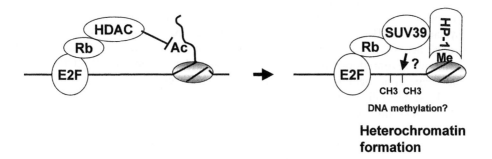

Fig. 4. Schematic of the deacetylase and methylase activities recruited to a promoter by Rb showing the potential sequential modifications on a promoter. Refs. [6,61].

sequence of events, mediated by Rb, which leads to chromatin condensation and repression of promoters. We know that E7 interacts with Rb and a complex containing HDAC activity through binding to Mi2-beta and the latter interaction is through the CR3 (zinc finger) domain [13]. We also know that expression of E7 will relieve the repression of Rb and this is probably achieved by competing for Rb binding with the HDAC complex as both bind in the same region of the B-pocket. However, since E7 can bind to Rb-independent repressive complexes through Mi2–beta interactions, E7 may also activate promoters other than those that are repressed by Rb. While the significance of these results remain unclear, the targeting of histone deacetylases provides yet another method by which E7 can de-repress gene transcription, and may explain the essential nature of the CR3 domain in activating E2F-regulated genes as well as immortalizing keratinocytes.

Effects of E7 on cyclin-CDKs

As previously discussed, viral genome amplification occurs in the differentiating epithelium, and, as such, HPV must force the cell to provide DNA replication components and activities in the absence of growth signals. The G1/S cyclins and cyclin-dependent kinases (cyclin D-cdk4, cyclin E-cdk2, and cyclin A-cdk2) are essential regulators of progression from the first gap phase through DNA synthesis [75,76]. In a proliferating cell, mitogenic stimulation signals the synthesis and assembly of cyclin D-cdk4 which contributes to inactivation of Rb via phosphorylation, leads to expression of cyclin E, and sequesters inhibitors of cdk2 (Cip/Kip inhibitors). Subsequently, cyclin E-cdk2 continues to inactivate Rb and phosphorylates other substrates essential for initiation of DNA replication and S phase entry. During S phase, cyclin A-cdk2 is assembled and remains active through the second gap phase. As integral components of the G1/S and S/G2 transitions, cyclin-cdks are logical targets of HPV. The effect of E7 on each of these cyclin-cdks is discussed below.

Cyclin E-cdk2 is essential for initiation of DNA synthesis and activates cell cycle progression even in the absence of cyclin D-cdk4 [28,43,63], underlining this complex as a potential target by which E7 propels cell cycle re-entry. Interestingly, several groups have demonstrated E7-mediated increases in cyclin E-cdk2 activity in asynchronous HFK [26,39,69,70]. In addition, E7-expressing cells maintain cyclin E-cdk2 activity in the presence of growth arrest signals such as epithelial differentiation, serum deprivation, and anchorage-independent growth [69,73]. This phenotype is abrogated by mutation of the CR1 region of E7, which is vital for cellular immortalization and transformation (see above). The mechanism by which E7 affects cyclin E-cdk2 activity *in vivo* has not been clearly defined, though several models have been proposed. The presence of E7 dysregulates cyclin E expression at the transcriptional (via relief of E2F-dependent repression) and post-transcriptional levels, suggesting that E7 may, in part, affect cyclin E-cdk2 activity via increased synthesis of cyclin E [11,50,81]. However, the contribution of elevated cyclin E expression is, alone, unlikely to be sufficient to increase cyclin E-cdk2 activity, because E7-expressing cells also exhibit dramatically elevated total and cdk2-associated levels of $p21^{CIP}$, a potent inhibitor of cdk2 [26,39]. High exogenous or endogenous cyclin E expression has been shown to be unable to maintain cyclin E-cdk2 activity in the presence of high levels of the Cip/Kip family of cdk inhibitors ($p21^{CIP}$ and $p27^{KIP}$, [1,74]).

E7 has been detected in complex with p107-cyclin E-cdk2, suggesting that E7 may affect cyclin E-cdk2 through physical interaction [54]. Indeed, recent evidence also suggests that E7 interacts with $p21^{CIP}$ and $p27^{KIP}$ *in vitro* and *in vivo*; and *in vitro* experiments have shown that E7 can bind to and derepress the kinase activity of $p21^{CIP}$-cyclin E-cdk2 complexes [26,39]. However, it is not clear whether active E7-$p21^{CIP}$-cyclin E-cdk2 complexes exist *in vivo*. Other reports refute the interaction between E7 and $p21^{CIP}$ [33,70] Our own research would concur with the latter findings and shows that E7 upregulates the level of $p21^{CIP}$, but that the cellular localization is altered in the presence of E7 with $p21^{CIP}$ found in the cytoplasm rather than the nucleus [79]. This is achieved by increase in Akt (protein kinase B) activity in the presence of E7 where Akt has been shown to phosphorylate $p21^{CIP}$ in the nuclear localization domain of the protein, causing accumulation in the cytoplasm [83]. An E7 mutant unable to bind Rb is defective in regulating Akt kinase activity, although the significance of this is unclear. Therefore, E7 indirectly down-regulates the ability of $p21^{CIP}$ to abrogate the cyclinE/cdk2 activity by altering the cellular localization so that it is confined to the cytoplasm and is in the wrong cellular compartment to effect cyclinE/cdk2 kinase activity.

Another mechanism by which E7 may activate cyclinE/cdk2 is through cdc25A. This protein is a phosphatase, which removes an inhibitory phosphate group from amino acid 15 on cdk2 [38], allowing it then to be activated by a cdk2-associated kinase (CAK) on a threonine residue at amino acid 160 [58]. E7 has been shown to up-regulate cdc25A at the transcriptional level by relieving Rb and HDAC repression [41,60] and that this regulation is independent of cell cycle progression [60]. This indicates that E7 is having a direct effect of the cdc25A promoter and it is not an indirect effect due to G1 to S-phase progression. Therefore E7 may regulate cyclinE/cdk2 activity through transcriptional up-regulation of: (1) cdc25A which will in turn activate cyclin E/cdk2 complexes; and (2) Akt activity which will cause the cytoplasmic localization of $p21^{CIP}$ and so prevent it from abrogating the activity of the cyclinE/cdk2 kinase complexes.

The effects of E7 on cyclin A-cdk2 (which regulates the progression of S phase) are similar to those observed for cyclin E-cdk2. Cyclin A-cdk2 activity is elevated in asynchronous HFK expressing E7 [50]. Deregulated expression of cyclin A may contribute to this phenotype, as the presence of E7 leads to constitutive expression of cyclin A throughout the cell cycle [50]. In concordance, E7-expressing fibroblasts synthesize cyclin A in the presence of growth arrest signals such as anchorage-independent suspension and serum starvation [73,81]. E7 has also been demonstrated to bind cyclin A-cdk2 in a complex involving E2F and p107 [4, 64]. The effect of E7 on the activities of the cyclin A-cdk2–E2F–p107 complex has not been elucidated. However, it is noteworthy that cyclin A-cdk2 inhibits the DNA-binding activity of E2F complexes as cells undergo DNA replication, and the expression of mutant E2F proteins lacking the cdk-binding site generates stable DNA-binding E2F complexes and arrests cells in S phase [42]. Because HPV requires the cellular DNA replication machinery for its own DNA synthesis, prolonging the time during which the terminally differentiating host cell is competent to synthesize DNA would benefit the replicative requirements of the virus. In concert, the effects of E7 on dysregulation of cyclin E-cdk2 (which regulates initiation of S phase) and cyclin A-cdk2 (which regulates the progression of S phase) may contribute to extending the duration of "replicative competence" in the host cell.

Effects of E7 on Rb stability

From the previous sections it is obvious that E7 targets Rb, and abrogates the ability of Rb to repression gene expression by remodeling chromatin so transcriptional activity is induced. However, a number of groups have observed that the expression of HPV-16, but not HPV-6 E7 causes a significant reduction in Rb protein levels in cells [9,29,32,40]. Degradation of Rb is proteasome mediated [9] and while the ability of E7 to bind to Rb is required, it is not sufficient for degradation as mutations in the CR1 domain bind Rb, but do not cause degradation [10,32,40]. In addition, there is also a region in the CR-2 domain, 3-prime to the Rb binding site on E7 that appears to be important for degradation [29]. While binding to Rb is necessary for the ability of E7 to induce Rb degradation and another site may be necessary for the actual process of degradation, there are other domains of E7 required for the abrogation of Rb function. For instance, the C-terminal domain of E7 is required for the ability of E7 to overcome cell cycle arrest and to immortalize human keratinocytes [13,37]. Therefore degradation of Rb, while seemingly necessary for E7 biology, is by itself not sufficient for the full array of E7 functions. This is further evidence that there are important biological properties of E7, which are Rb-independent. These functions may include the ability of E7 to bind to complexes containing histone deacetylase activity independent of Rb, or other interactions, which have yet to be uncovered. All of the published experiments on Rb degradation have been carried out with E7 alone. However, in the presence of E6, the degradation of Rb is reduced (Baglia and McCance, unpublished data), and since both proteins are present in the natural infection, the significance of Rb degradation is unclear.

Summary

Papillomaviruses, which produce either benign or malignant lesions have to replicate in cells that are programmed to terminally differentiate. Therefore, both groups of viruses have to stimulate cells into S-phase for successful replication. HPV-16 E7 stimulates G1 to S-phase progression by a series of interactions, which hinge on the ability of the protein to modulate cellular transcription, usually of genes that are essential for successful passage through G1 and into S-phase, such as cdc25A and cyclin E/cdk2. In addition, E7 can regulate the activity of kinase inhibitors such as p21^{CIP}, so that they are shunted to the cytoplasm and away from the site of action of cyclin kinases in the nucleus. The mechanism involves the activity of the Akt family of kinases, but how this is regulated by E7 is unclear, although it may involve genes modulated by Rb. While the interaction with Rb is important, there are Rb-independent interactions, which are also important for E7 biology and it will be interesting in the coming years to determine the nature of these proteins and their function.

While progress has been made in delineating the pathways used by malignant viruses to stimulate the cell cycle, the big unknown is how HPV-6 and other benign viruses achieve the same feat. So far, nearly all the biology observed with HPV-16 E7 is not observed with HPV-6 E7, and so it is difficult to see how these latter viruses could stimulate G1 to S-phase by the same pathways. The next few years will I hope see some advances in understanding the biology of the low risk types.

Acknowledgements

I would like to thank past and present members of the laboratory for their contributions to some of the research described here. The research described was funded by the National Institutes of Health, USA.

References

1. Alevizopoulos, K, Catarin, B, Vlach, J, Amati B. A novel function of adenovirus E1A is required to overcome growth arrest by the CDK2 inhibitor p27(Kip1) EMBO J 1998; 17: 5987–59897.
2. Angel P, Karin M. The role of Jun, Fos and the AP-1 complex in cell-proliferation and transformation. Biochim Biophys Acta 1991; 1072: 129–157.
3. Antinore MJ, Birrer MJ, Patel D, Nader L, McCance DJ. The human papillomavirus type 16 E7 gene product interacts with and trans-activates the AP1 family of transcription factors EMBO J 1996; 15: 1950–1960.
4. Arroyo M, Bagchi S, Raychaudhuri P. Association of the human papillomavirus type 16 E7 protein with the S-phase-specific E2F-cyclin A complex. Molec Cell Biol 1993; 13: 6537–6546.
5. Banks L, Edmonds C, Vousden K. Ability of the HPV16 E7 protein to bind RB and induce DNA synthesis is not sufficient for efficient transforming activity in NIH3T3 cells. Oncogene 1990; 5: 1383–1389.
6. Bannister AJ, Zegerman P, Partridge JF, Miska EA, Thomas JO, Allshire RC, Kouzarides T. Selective recognition of methylated lysine 9 on histone H3 by the HP1 chromo domain. Nature 2001; 410: 120–124.
7. Barbosa MS, Edmonds C, Fisher C, Schiller JT, Lowy DR, Vousden KH. The region of the HPV E7 oncoprotein homologous to adenovirus E1a and Sv40 large T antigen contains separate domains for Rb binding and casein kinase II phosphorylation. EMBO J 1990; 9: 153–160.
8. Barbosa MS, Schlegel R. The E6 and E7 genes of HPV-18 are sufficient for inducing two-stage *in vitro* transformation of human keratinocytes. Oncogene 1989; 4: 1529–1532.

9. Berezutskaya E, Bagachi S. The human papillomavirus E7 oncoprotein functionally interacts with the S4 subunit of the 26S proteasome. J Biol Chem 1997; 272: 30135–30140.

10. Berezutskaya E, Yu B, Morozov A, Raychaudhuri P, Bagchi S. Differential regulation of the pocket domains of the retinoblastoma family proteins by the HPV16 E7 oncoprotein. Cell Growth Differ 1997; 8: 1277–1286.

11. Botz J, Zerfass-Thome K, Spitkovsky D, Delius H, Vogt B, Eilers M, Hatzigeorgiou A, Jansen-Durr P. Cell cycle regulation of the murine cyclin E gene depends on an E2F binding site in the promoter. Molec Cell Biol 1996; 16: 3401–3409.

12. Brehm A, Miska EA, McCance DJ, Reid JL, Bannister AJ, Kouzarides T. Retino-blastoma protein recruits histone deacetylase to repress transcription [see comments]. Nature 1998; 391: 597–601.

13. Brehm A, Nielsen SJ, Miska EA, McCance DJ, Reid JL, Bannister AJ, Kouzarides T. The E7 protein associates with Mi2 and histone deacetylase activity to promote cell growth. EMBO J 1999; 18: 2449–2458.

14. Brokaw JL, Yee CL, Munger K. A mutational analysis of the amino terminal domain of the human papillomavirus type 16 E7 oncoprotein. Virology 1994; 205: 603–607.

15. Chesters PM, Vousden KH, Edmonds C, McCance DJ. Analysis of human papilloma-virus type 16 open reading frame E7 immortalizing function in rat embryo fibroblast cells. J Gen Virol 1990; 71: 449–453.

16. Davies R, Hicks R, Crook T, Morris J, Vousden K. Human papillomavirus type 16 E7 associates with a histone H1 kinase and with p107 through sequences necessary for transformation. J Virol 1993; 67: 2521–2528.

17. Defeo-Jones D, Vuocolo GA, Haskell KM, Hanobik MG, Kiefer DM, McAvoy EM, Ivey-Hoyle M, Brandsma JL, Oliff A, Jones RE. Papillomavirus E7 protein binding to the retinoblastoma protein is not required for viral induction of warts. J Virol 1993; 67: 716–725.

18. Demers GW, Espling E, Harry JB, Etscheid BG, Galloway DA. Abrogation of growth arrest signals by human papillomavirus type 16 E7 is mediated by sequences required for transformation. J Virol 1996; 70: 6862–6869.

19. Dowdy SF, Hinds PW, Louie K, Reed SI, Arnold A, Weinberg RA. Physical interaction of the retinoblastoma protein with human D cyclins. Cell 1993; 73: 499–511.

20. Dunaief JL, Strober BE, Guha S, Khavari PA, Alin K, Luban J, Begemann M, Crabtree GR, Goff SP. The retinoblastoma protein and BRG1 form a complex and cooperate to induce cell cycle arrest. Cell 1994; 79.

21. Dyson N. The regulation of E2F by Rb-family proteins. Genes & Devel 1998; 12: 2245–2262.

22. Dyson N, Guida P, Munger K, Harlow E. Homologous sequences in adenovirus E1A and human papillomavirus E7 proteins mediate interaction with the same set of cellular proteins. J Virol 1992; 66: 6893–6902.

23. Dyson N, Howley PM, Munger K, Harlow E. The human papilloma virus-16 E7 onco-protein is able to bind to the retinoblastoma gene product. Science 1989; 243: 934–937.

24. Eckert RL, Welter JF. Transcription factor regulation of epidermal keratinocyte gene expression. Molec Biol Rep 1996; 23: 59–70.

25. Edmonds C, Vousden KH. A point mutational analysis of human papillomavirus type 16 E7 protein. J Virol 1989; 63: 2650–2656.

26. Funk JO, Waga S, Harry JB, Espling E, Stillman B, Galloway DA. Inhibition of CDK activity and PCNA-dependent DNA replication by p21 is blocked by interaction with the HPV-16 E7 oncoprotein. Genes & Devel 1997; 11: 2090–2100.

27. Gage JR, Meyers C, Wettstein FO. The E7 proteins of the nononcogenic human papillomavirus type 6b (HPV-6b) and of the oncogenic HPV-16 differ in retinoblastoma protein binding and other properties. J Virol 1990; 64: 723–730.

28. Geng Y, Whoriskey W, Park MY, Bronson RT, Medema RH, Li T, Weinberg RA, Sicinski P. Rescue of cyclin D1 deficiency by knockin cyclin E. Cell 1999; 97: 767–777.

29. Giarre M, Caldeira S, Malanchi I, Ciccolini F, Joao Leao M, Tommasino M. Induction of pRb degradation by human papillomavirus type 16 E7 protein is essential to efficiently overcome p16^{INK4a}-imposed G1 cell cycle arrest. J Virol 2001; 75: 4705–4712.

30. Halbert CL, Demers GW, Galloway DA. The E6 and E7 genes of human papillomavirus type 6 have weak immortalizing activity in human epithelial cells. J Virol 1992; 66: 2125–2134.

31. Heck DV, Yee CL, Howley PM, Munger K. Efficiency of binding the retinoblastoma protein correlates with the transforming capacity of the E7 oncoproteins of the human papillomaviruses. Proc Natl Acad Sci USA 1992; 89: 4442–4446.

32. Helt AM, Galloway DA. Destabilization of the retinoblastoma tumor suppressor by human papillomavirus type 16 E7 is not sufficient to overcome cell cycle arrest in human keratinocytes. J Virol 2001; 75: 6737–6747.

33. Hickman ES, Bates S, Vousden KH. Pertubation of the p53 response by human papillomavirus type 16 E7. J Virol 1997; 71: 3710–3718.

34. Huang PS, Patrick DR, Edwards G, Goodhart PJ, Huber HE, Miles L, Garsky VM, Oliff A, Heimbrook DC. Protein domains governing interactions between E2F, the retinoblastoma gene product, and human papillomavirus type 16 E7 protein. Molec Cell Biol 1993; 13: 953–960.

35. Hudson JB, Bedell MA, McCance DJ, Laiminis LA. Immortalization and altered differentiation of human keratinocytes *in vitro* by the E6 and E7 open reading frames of human papillomavirus type 18. J Virol 1990; 64: 519–526.

36. Jenuwein T, Allis CD. Translating the histone code. Science 2001; 293: 1074–1080.

37. Jewers RJ, Hildebrandt P, Ludlow JW, Kell B, McCance DJ. Regions of human papillomavirus type 16 E7 oncoprotein required for immortalization of human keratinocytes. J Virol 1992; 66: 1329–1335.

38. Jinno S, Suto K, Nagata A, Igarashi M, Kanaoka Y, Nojima H, Okayama H. cdc25A is a novel phosphatase functioning early in the cell cycle. EMBO J 1994; 13: 1549–1556.

39. Jones DL, Alani RM, Munger K. The human papillomavirus E7 oncoprotein can uncouple cellular differentiation and proliferation in human keratinocytes by abrogating p21Cip1-mediated inhibition of cdk2. Genes & Devel 1997; 11: 2101–2111.

40. Jones DL, Thompson DA, Munger K. Destabilization of the RB tumor suppressor protein and stabilization of p53 contribute to HPV type 16 E7-induced apoptosis. Virology 1997; 239: 97–107.

41. Katich SC, Zerfass-Thome K, Hoffmann I. Regulation of the Cdc25A gene by the human papillomavirus Type 16 E7 oncogene. Oncogene 2001; 20: 543–550.

42. Krek W, Xu G, Livingston DM. Cyclin A-kinase regulation of E2F-1 DNA binding function underlies suppression of an S phase checkpoint. Cell 1995; 83: 1149–1158.

43. Krude T, Jackman M, Pines J, Laskey RA. Cyclin/cdk-dependent initiation of DNA

replication in a human cell-free system. Cell 1997; 88: 109–119.

44. Lai A, Kennedy BK, Barbie DA, Bertos NR, Yang XJ, Theberge MC, Tsai SC, Seto E, Zhang Y, Kuzmichev A, Lane WS, Reinberg D, Harlow E, Branton PE. RBP1 recruits the mSIN3-histone deacetylase complex to the pocket of retinoblastoma tumor suppressor family proteins found in limited discrete regions of the nucleus at growth arrest. Molec Cell Biol 2001; 21: 2918–2932.

45. Lee J.-O, Russo AA, Pavletich NP. Structure of the retinoblastoma tumor-suppressor pocket domain bound to a peptide from HPV E7. Nature 1998; 391: 859–865.

46. Ludlow JW, DeCaprio JA, Huang CM, Lee WH, Paucha E, Livingston DM. SV40 large T antigen binds preferentially to an underphosphorylated member of the retinoblastoma susceptibility gene product family. Cell 1989; 56: 57–65.

47. Ludlow JW, Shon J, Pipas JM, Livingston DM, DeCaprio JA. The retinoblastoma susceptibility gene product undergoes cell cycle-dependent dephosphorylation and binding to and release from SV40 large T. Cell 1990; 60: 387–396.

48. Magnaghi-Jaulin L, Groisman R, Naguibneva I, Robin P, Lorain S, Le Villain JP, Troalen F, Trouche D, Harel-Bellan A. Retinoblastoma protein represses transcription by recruiting a histone deacetylase [see comments]. Nature 1998; 391: 601–605.

49. Marks PA, Rifkind RA, Richon VM, Breslow R, Miller T, Kelly WK. Histone deacetylases and cancer: causes and therapies. Nature Rev: Cancer 2001; 1: 194–202.

50. Martin LG, Demers GW, Galloway DA. Disruption of the G1/S transition in human papillomavirus type 16 E7-expressing human cells is associated with altered regulation of cyclin E. J Virol 1998; 72: 975–985.

51. Massimi P, Banks L. Differential phosphorylation of the HPV-16 E7 oncoprotein during the cell cycle. Virology 2000; 276: 388–394.

52. Mazzarelli JM, Atkins GB, Geisberg JV, Ricciardi RP. The viral oncoproteins Ad5 E1A, HPV16 E7 and SV40 TAg bind a common region of the TBP-associated factor-110. Oncogene 1995; 11: 1859–1864.

53. McCance DJ, Kopan R, Fuchs E, Laimins LA. Human papillomavirus type 16 alters human epithelial cell differentiation *in vitro*. Proc Natl Acad Sci USA 1988; 85: 7169–7173.

54. McIntyre MC, Frattini MG, Grossman SR, Laimins LA. Human papillomavirus type 18 E7 protein requires intact Cys-X-X-Cys motifs for zinc binding, dimerization, and transformation but not for Rb binding. J Virol 1993; 67: 3142–3150.

55. Munger K, Phelps WC. The human papillomavirus E7 protein as a transforming and transactivating factor. Biochim Biophys Acta 1993; 1155: 111–123.

56. Munger K, Werness BA, Dyson N, Phelps WC, Harlow E, Howley PM. Complex formation of human papillomavirus E7 proteins with the retinoblastoma tumor suppressor gene product. EMBO J 1989; 8: 4099–4105.

57. Munger K, Yee CL, Phelps WC, Pietenpol JA, Moses HL, Howley PM. Biochemical and biological differences between E7 oncoproteins of the high- and low-risk human papillomavirus types are determined by amino-terminal sequences. J Virol 1991; 65: 3943–3948.

58. Nagahara H, Ezhevsky SA, Vocero-Akbani AM, Kaldis P, Solomon MJ, Dowdy SF. Transforming growth factor beta targeted inactivation of cyclin E: cyclin-dependent kinase 2 (Cdk2) complexes by inhibition of Cdk2 activating kinase activity. Proc Natl Acad Sci USA 1999; 96: 14961–14966.

59. Nead MA, Baglia LA, Antinore MJ, Ludlow JW, McCance DJ. Rb binds c-Jun and activates transcription. EMBO J 1998; 17: 2342–2352.

60. Nguyen DX, Westbrook TF, McCance DJ, HPV-16 E7 maintains elevated levels of the cdc25A tyrosine phosphatase during deregulation of cell cycle arrest. J Virol 2002; 76: 619–632.

61. Nielsen SJ, Schneider R, Bauer UM, Bannister AJ, Morrison A, O'Carroll D, Firestein R, Cleary M, Jenuwein T, Herrera RE, Kouzarides T. Rb targets histone H3 methylation and HP1 to promoters. Nature 2001; 412: 561–565.

62. Nishitani J, Nishinaka T, Cheng CH, Rong W, Yokoyama KK, Chiu R. Recruitment of the retinoblastoma protein to c-Jun enhances transcription activity mediated through the AP-1 binding site. J Biol Chem 1999; 274: 5454–5461.

63. Ohtsubo M, Theodoras AM, Schumacher J, Roberts JM, Pagano M. Human cyclin E, a nuclear protein essential for the G1-to-S phase transition. Molec & Cell Biol 1995; 15: 2612–2624.

64. Pagano M, Pepperkok R, Verde F, Ansorge W, Draetta G. Cyclin A is required at two points in the human cell cycle. EMBO J 1992; 11: 961–971.

65. Patrick DR, Oliff A, Heimbrook DC. Identification of a novel retinoblastoma gene product binding site on human papillomavirus type 16 E7 protein. J Biol Chem 1994; 269: 6842–6850.

66. Phelps WC, Munger K, Yee CL, Barnes JA, Howley PM. Structure-function analysis of the human papillomavirus type 16 E7 oncoprotein. J Virol 1992; 66: 2418–2427.

67. Phillips AC, Vousden KH. Analysis of the interaction between human papillomavirus type 16 E7 and the TATA-binding protein, TBP. J Gen Virol 1997; 78: 905–909.

68. Pirisi L, Yasumoto S, Feller M, Doniger J, DiPaolo JA. Transformation of human fibroblasts and keratinocytes with human papillomavirus type 16 DNA. J Virol 1987; 61: 1061–1066.

69. Ruesch M, Laimins LA. Human papillomavirus oncoproteins alter differentiation-dependent cell cycle exit on suspension in semisolid medium. Virology 1998; 250: 19–29.

70. Ruesch MN, Laimins LA. Initiation of DNA synthesis by human papillomavirus E7 oncoproteins is resistant to p21-mediated inhibition of cyclin E-cdk2 activity. J Virol 1997; 71: 5570–5578.

71. Sang BC, Barbosa MS. Single amino acid substitutions in "low-risk" human papillomavirus (HPV) type 6 E7 protein enhance features characteristic of the "high-risk" HPV E7 oncoproteins. Proc Natl Acad Sci USA 1992; 89: 8063–8067.

72. Schmitt A, Harry JB, Rapp B, Wettstein FO, Iftner T. Comparison of the properties of the E6 and E7 genes of low- and high-risk cutaneous papillomaviruses reveals strongly transforming and high Rb-binding activity for the E7 protein of the low-risk human papillomavirus type 1. J Virol 1994; 68: 7051–7059.

73. Schulze A, Mannhardt B, Zerfass-Thome K, Zwerschke W, Jansen-Durr P. Anchorage-independent transcription of the cyclin A gene induced by the E7 oncoprotein of human papillomavirus type 16. J Virol 1998; 72: 2323–2334.

74. Sewing A, Wiseman B, Lloyd AC, Land H. High-intensity Raf signal causes cell cycle arrest mediated by p21Cip1. Molec Cell Biol 1997; 17: 5588–5597.

75. Sherr CJ. Mammalian G1 cyclins. Cell 1993; 73: 1059–1065.

76. Sherr CJ, Roberts JM. CDK inhibitors: positive and negative regulators of G1-phase progression. Genes & Devel 1999; 13: 1501–1512.

77. Takami Y, Sasagawa T, Sudiro TM, Yutsudo M, Hakura A. Determination of the functional difference between human papillomavirus type 6 and 16 E7 proteins by their 30 N-terminal amino acid residues. Virology 1992; 186: 489–495.

78. Tamaru H, Selkar EU. A histone H3 methyltransferase controls DNA methylation in *Neurospora crassa*. Nature 2001; 414: 277–283.

79. Westbrook TF, Nguyen D, Thrash B, McCance DJ. 2002. E7 abrogates RAF-induced arrest by mislocalization of p21^{CIP1}. Mol Cell Biol; in press.

80. Wu EW, Clemens KE, Heck DV, Munger K. The human papillomavirus E7 oncoprotein and the cellular transcription factor E2F bind to separate sites on the retinoblastoma tumor suppressor protein. J Virol 1993; 67: 2402–2407.

81. Zerfass K, Schulze A, Spitkovsky D, Friedman V, Henglein B, Jansen-Durr P. Sequential activation of cyclin E and cyclin A gene expression by human papillomavirus type 16 E7 through sequences necessary for transformation. J Virol 1995; 69: 6389–6399.

82. Zhang Y, LeRoy G, Seelig HP, Lane WS, Reinberg D. The dermatomyositis-specific autoantigen Mi2 is a component of a complex containing histone deacetylase and nucleosome remodeling activities. Cell 1998; 95: 279–289.

83. Zhou BP, Liao Y, Xia W, Spohn B, Lee M.-H, Hung M.-C. Cytoplasmic localization of p21$^{CIP1/WAF1}$ by Akt-induced phosphorylation in *HER-2*/neu-overexpressing cells. Nature Cell Biol 2001; 3: 245–252.

Human Papillomaviruses
D.J. McCance (editor)

Biology of the E4 protein

Sally Roberts
Cancer Research Campaign Institute for Cancer Studies, The Medical School, University of Birmingham, Birmingham B15 2TA, UK

Introduction

Papillomavirus (PV) genome amplification, capsid expression and assembly of progeny are late virus events triggered in cells that have left the basal cell layer and begun to migrate up into the differentiating layers. This productive phase of the life-cycle is accompanied by expression of the E4 gene product, an early protein by virtue of the position of the gene in the virus genome (the E4 open-reading frame (ORF) overlaps in its entirety with the E2 ORF), but regarded as an intermediate or late protein by the pattern of expression in warts and tumours. The E4 gene product was identified in the late 1980s, and shown neither to have cellular transforming properties nor an intimate role in the establishment of virus episomes. Neither did it form part of the virus capsid. Understandably therefore, those viral proteins involved in these aspects of papillomavirus biology received much more attention than E4, and the function of E4 in the virus life-cycle has remained an enigma. Furthermore, the technical difficulties associated with reproduction of the normal pattern of epithelial terminal differentiation in cultured epithelial cells have seriously hampered efforts to elucidate E4's role. Nevertheless, E4 represents an interesting challenge to the virologist and a number of functions have been proposed. These include: inhibition of terminal differentiation to preserve cellular integrity and enhance virus synthesis [73]; involvement in vegetative viral DNA replication [6,45], perhaps by interfering with the viral or cellular mechanisms that inhibit virus replication in basal cells [6]; control of virus maturation [21,61], and mediation of virus release from the epithelial squames by disruption of the cytoskeleton [24]. In more recent years, identification of cellular factors and pathways that are targeted by E4 has shed more light on how E4 interacts with the host cell. These studies show that there are differences in E4 activities between individual virus types and this may reflect differences in the biology of the viruses. Therefore, although E4 might perform a common role in the virus's productive life-cycle, this could be achieved through different, but perhaps overlapping, mechanisms. Another important outcome of these studies has been the assignment of particular biological and biophysical properties to specific regions of the E4 protein. The emerging picture shows that, despite their sequence diversity, the regions of E4 that mediate these activities correspond to specific sequence motifs and regions of homology between

E4 proteins. It is also suggested that E4 could perform multiple functions during the virus life-cycle, partly through generation of a series of multiple E4 species that arise by protein modification as the infected cell moves up through the differentiating lesion. This chapter will concentrate on a discussion of the recent advances in E4 research and the conclusions about E4 function that have been drawn from these studies, but first it will be important to describe our current knowledge of E4 expression *in vivo*.

E4 expression and cellular distribution in naturally-occurring warts

The E4 protein is in fact a spliced product E1 ^ E4 formed between the first five amino acids of E1 and the E4 ORF. A number of viral polycistronic messenger RNA transcripts contain the E1 ^ E4 ORF. The most abundant class (E1 ^ E4, E5) has been described for cutaneous and mucosal HPV types and is derived from a differentiation-inducible promoter in the E7 ORF [15,38,42,56,62,82]. The other E1 ^ E4 ORF-containing messages have the potential to also express the major and minor capsid proteins, L1 and L2 [8,15,62,82] (and see Chapter 3). Throughout this chapter the E1 ^ E4 protein will be referred to as E4. There is a striking correlation between the appearance of E4 protein in differentiating cells of the suprabasal layers and the onset of vegetative PV DNA replication [2,6,26,58,61,81]. In human papillomavirus type 16 (HPV16)-positive tumours this occurs in the mid-spinous layers whilst in cutaneous warts, for example those produced by HPV1 and HPV63 infection, these two events occur much deeper in the differentiating epithelia and are observed as early as the parabasal layers [6,26]. E4 is also present in superficial cells that express the structural proteins L1 and L2 [6,7,17,21,22,26]. In some PV-induced tumours E4 expression precedes initiation of capsid expression, for example, in HPV63 infections E4 protein expression is triggered in cells of the parabasal layers, but L1 protein is not detected until the cells have reached the granular layer [26]. In others, for example HPV11, it parallels capsid expression [7]. Thus, while cells that are positive for E4 are not necessarily positive for capsid proteins the converse is not true. Interestingly, in HPV1 warts L1 protein is detected in the lower spinous cells along with E4, while L2 protein is restricted to highly differentiating cells [29]. This divergent expression of the capsid proteins is supported by the distribution of viral transcripts [29]. Although there appear to be differences in the level of control of late gene expression between individual PV types, overall, E4 expression *in vivo* is linked to the productive phase of the virus life-cycle, and this is supported by *in vitro* studies [45,75]. For example, keratinocytes containing episomal copies of HPV31 DNA, when induced to differentiate by culturing in methylcellulose switched to amplification of the viral genome and induction of E4 protein in a coordinated manner [75]. This study also demonstrated that while early markers of differentiation such as involucrin and transglutaminase were compatible with E4 expression, expression of the differentiation-specific keratins, and the granular layer specific protein filaggrin, was not necessary [75]. In fact, several reports have shown that induction of E4 expression is not necessarily dependent on the infected cell undergoing terminal differentiation. Growth-arrest of mouse C127 cells transformed with bovine PV type 1 (BPV1) was sufficient to induce a switch between early stage BPV1 genome replication to late stage amplification and the majority of these cells expressed E4 [45]. In canine oral (CO) PV-induced lesions, E4 expression and viral genome amplification occur in the same cells of the non-differentiating basal epithelial layers [58]. These studies indicate that induction of E4 expression and replication of viral DNA are interdependent processes, but are not always triggered by epithelial terminal differentiation. It is important to note that the level of virus yield correlates with the level of E4 expression [6,21,22]. This is most aptly illustrated by HPV2, as warts induced by this virus are extremely variable in virus production, but high E4 content correlates with production of large amounts of virus and *vice versa* [22]. Taken together, the studies of E4 expression in warts and tumours suggests that E4 might be necessary for efficient generation of progeny virus, perhaps acting at several stages of the productive programme.

The E4 gene products have a primarily cytoplasmic location, but have been observed in the nuclei of lesions caused by both cutaneous and mucosal virus types (Table 1). Intracellular distribution of E4 does differ between viruses, and it is apparent that the variation is related, to some extent, to the epithelial tropism of the virus. In cutaneous benign lesions, E4 assembles into inclusion bodies whose ultra-structural appearance and size is quite variable between the different cutaneous viruses [16,22]. For example, in HPV1 warts they appear in lower suprabasal layers as numerous irregularly shaped, homogenously electron dense bodies, but coalesce into larger structures in cells of the upper layers. In contrast, the HPV4 E4 protein is associated with a single inclusion with a fine fibrillar structure that progressively fills the cytoplasm and distorts the nucleus, forming the typical "signet ring" cell. On the other hand, distribution of mucosa-specific E4 proteins is much more variable between members of this group (Table 1). In pre-malignant tumours, HPV16 E4 associates with cytoplasmic filamentous networks, most likely the keratin inter-mediate filaments (IFs) [26,79, and our unpublished data], and in superficial cells is detected in a small inclusion-like body adjacent to the nucleus [26]. However, in another study HPV16 E4 was shown to be exclusively nuclear in superficial cells, localising to 25–35 nm spherical structures [61]. In organotypic raft cultures of an HPV31b-positive keratinocyte line, CIN-612, E4 forms a punctate pattern through-out the cytoplasm of suprabasal cells [66], and in differentiating cells of an HPV11-infected human foreskin implant it accumulates at the cell boundary [7]. Whilst some of the observed differences in E4 distribution between HPVs may reflect the use of different tissue fixation methodologies and specificity of antibodies used, it would seem that E4 localization is influenced by its epithelial site of expression. Thus, differences in the biology of cutaneous vs. mucosa epithelia and/or differences in the biology of individual virus types, i.e. benign vs. oncogenic may reflect variation in E4 subcellular localization. To put this another way, E4 might perform a common role in the virus life-cycle, but the mechanism of action of this viral protein may be different between virus types, and this could be related to their tropism and/or pathogenicity.

Table 1

Properties of PV E4 proteins expressed in PV-infected tissues and cultured epithelial cells

Virus type	Polypeptides expressed in vivo[a]	Intracellular localization in vivo	Modifications	IF-associated protein	Refs.
HPV1	10/11K, 16/17K, 21/23K, 32/34K and higher molecular weight species	Multiple cytoplasmic/nuclear inclusions and diffuse cytoplasmic distribution	Phosphorylation on serine and threonine residues. Phosphorylated by PKA in vitro. Binds divalent metal ion zinc.	Yes, but E4-IF networks do not collapse	4, 6, 16, 21, 37, 68, 71
HPV2	16.5/18K, 33K and minor species of 22K/25K	Multiple cytoplasmic (nuclear?) inclusions	ND	Yes, and induces collapse of keratin networks	5, 22
HPV4	20/21K, 40K and minor species of 15.5K/16.5K	Single, large cytoplasmic inclusion	ND	ND	22
HPV6/11	10/11K	ND. Cytoplasmic, at cell periphery in HPV11-infected human foreskin	Phosphorylation on serine and threonine residues. Phosphorylated by PKA and MAPK in vitro	Likely, filamentous E4 networks are formed in monolayer epithelial cells	7, 12, 87
HPV16	ND. In epithelial cells one study identified a 10K protein. Another, polypeptides of 11K, 13K, 26K, 36K	Cytoplasmic filamentous, also single juxtanuclear inclusion in superficial cells. Nuclear structures (25–35 nm)	ND	Yes, and induces collapse of keratin networks	24, 26, 61, 69, 70, 79
HPV31	11K, 18K, 22K, 33K, 44K. In monolayer cultures, 8K and 11K	ND, but in organotypic raft cultures of HPV31b cell line, CIN 612-9E–cytoplasmic punctate staining	ND	Yes, and induces collapse of keratin networks	66
BPV1	16K, 21K, 30K, 42K	ND; but cytoplasmic perinuclear in BPV1 transformed C127 mouse cells	ND	ND	45

Virus type	Polypeptides expressed *in vivo*[a]	Intracellular localization *in vivo*	Modifications	IF-associated protein	Refs.
BPV4	ND	Cytoplasmic	ND	Filamentous E4 structures are formed in monolayer epithelial cells	2
COPV	ND	Cytoplasmic inclusions plus a diffuse distribution. Nuclear, at periphery of nucleoli	ND	ND	58

[a]Where known the size of the full-length E4 polypeptide is underlined. IF, keratin intermediate filaments; ND, not determined; HPV, human papillomavirus; BPV, bovine papillomavirus; COPV, canine oral papillomavirus.

Structure of E4 proteins

The E4 gene is particularly varied in length and amino acid composition between virus types. However, there is an apparent retention of sequence characteristics such as stretches of proline residues and charged amino acids, and the predominance of serine and threonine between some virus types. For details of alignments of E4 gene and protein sequences, the reader should refer to the review by Doorbar and Myers [28]. Although the overall homology between E4 proteins is not great, there is appreciable conservation of primary sequence at the N and C termini, with the central domain of the protein showing maximum diversity [25,68,70,92]. C-terminal homology is found between those viruses that have a similar site of infection, i.e. skin or anogenital epithelium, and thus the role of this region in E4 function may be linked to the tropism (and/or pathogenicity) of the virus [68, 70]. In contrast, a cluster of leucines (LLXLL) near to the N-terminus is found in viruses that infect cutaneous or mucosal epithelia [25,68], although this motif is not ubiquitous to all E4s. The extreme N-terminus corresponds to the small exon derived from the N-terminus of the E1 ORF, and this is also moderately conserved between different viruses. Perhaps not surprisingly, the emerging profiles of activity associated with E4 are being mapped to sequences corresponding to the conserved regions and motifs largely localized to the N and C-terminal domains of the molecule (Fig. 1). It has been suggested that perhaps the more diverse central region of E4 acts as a hinge or flexible linker separating the "functional" N and C-terminal domains [70]. The E4 ORF overlaps the region of the E2 predicted to form a flexible linker between the transactivating and DNA binding domains, itself variable in sequence content and length. Sequence analysis of this region in a group of HPV16 intratypic variants showed that far more amino acid variation was apparent in the E2 hinge than in the E4 ORF, and furthermore, non-conservative changes in E4 occurred outside those regions suggested to be important in E4 biology [30]. This study indicates that there

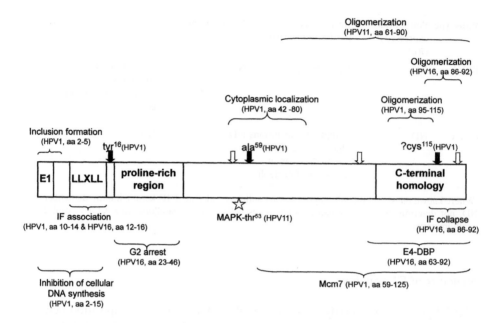

Fig. 1. Schematic representation of the structure of E4 is shown with the suggested sites of post-translational modifications and regions important in the identified E4 activities. HPV type and amino acids involved is given in brackets. Shaded arrows indicate the position of protease cleavage sites within HPV1 E4. The position of a putative cleavage site at the C-terminus (cys[115]) of HPV1 E4 was based on the reactivity of peptide-specific antibodies. Open arrows show the position of histidine amino acids in HPV1 E4 that bind the divalent metal ion, zinc. Assembled from data given in Refs. [4,11,12,19,23,68,70–72,74].

may be a selective pressure to maintain specific E4 sequences and supports the notion that the N and C terminal sequences are important in E4 structure and function [68,70].

No clues to E4 function can be determined from homologies with cellular components or other viral proteins. Only the HPV1 E4 protein has been found to have significant homology to a cellular factor—this is with the domain of human skeletal actin that is involved in self-interactions and interactions with myosin [68].

Posttranslational E4 modification

The E4 protein is modified through a series of posttranslational events including N-terminal protein cleavage, phosphorylation, and self-multimerization (Table 1). Much of the detail regarding the nature of these modifications has come from the study of the HPV1 E4 protein, largely as a result of its abundance in warts (E4 constitutes up to 30% of total wart protein [6,21]). In HPV1 warts at least eight prominent species of E4 are detected as protein doublets of 10/11K, 16/17K, 21/23K and 32/34K [6,21]. This family of polypeptides is further expanded by phosphorylation at multiple serines and threonine residues throughout the E4 molecule [6,37]. The 17K polypeptide represents the full-length E1 ^ E4 molecule of 125 amino acids, and the 34K species the dimeric form of 17K. The smaller species are derived from the full-length molecule through limited proteolytic cleavage. Conversion to 16K requires the removal of the extreme N-terminal 15 amino acids to tyrosine[16] [25,68]. Further proteolysis generates the 10 and 11K proteins [25], with alanine[59] as the N-terminal residue [68], and these species may have also lost extreme C-terminal residues [25] (Fig. 1). The minor species of 21K is a heterodimer of the 10 and 11K species and 23K is most probably formed by cleavage of the oligomeric forms [25,27]. The cleavage sites in the E4 molecule are not part of known protease recognition sites and a protease that cleaves these proteins has not been identified. It is unlikely that E4 proteolysis is carried out by a protease specific only to terminally-differentiating keratinocytes, because HPV1 E4 expressed in insect Sf9 cells is cleaved into similar-sized products found *in vivo* [71]. An intriguing possibility is that the E4 primary sequence encodes a protease activity and thus cleaves itself. Chemical cross-linking studies of E4 purified from warts or a baculovirus expression system demonstrated that E4 also assembles into larger complexes representing trimers and/or tetramers and higher orders of complexity [4,27]. These complexes are not dependent on disulphide linkages for their formation and yet are remarkably stable, persisting even in high concentrations (6 M) of urea [3,4,71]. Phosphorylation may promote the formation of HPV1 E4 oligomers [37] and E4-bound zinc ions may influence their stability [3,71]. At this stage of understanding HPV1 E4 structure and function it is not known if the smaller oligomers represent distinct functional forms of E4 or if they are intermediates in the assembly of the E4 inclusion bodies. Sequences in the C-terminal half of the HPV1 protein mediate self-multimerization [27] and genetic analysis identified a stretch of amino acids between residues 95 and 115 as essential for oligomerization [4]. Hydrophobic residues within this region are of particular importance [4]. Interestingly, they dominate one face of an amphipathic helix that this region is predicted to adopt, suggesting that E4:E4 associations might be mediated through interactions between hydrophobic faces of α-helical structures [4]. Whatever the precise nature of HPV1 E4 self-interactions, it is probable that other cutaneous E4s that share homology with HPV1 E4's C-terminal region have a similar mode of self-association [4,68].

Less is known about modification of E4 proteins of other virus types, although they too exist as multiple forms in naturally-occurring tumours and in tissue culture expression systems (Table 1) suggesting that they may also be modified by proteolysis and self-multimerization. The HPV11 E4 protein has been shown to self-associate [11] and exists as a phosphoprotein *in vivo* [12] (Table 1). Interestingly, it appears that the same region of the E4 molecule implicated in HPV1 multimerization, the C-terminus, is also important for self-association between E4's of anogenital types. Mutational analysis of the HPV11 and HPV16 proteins identified a stretch of amino acids at the C-terminus that was necessary for oligomerization [11,70] (Fig. 1), and this region is homologous between E4 proteins of other anogenital virus types [70].

Modification of the E4 proteins varies as the infected keratinocyte moves up through the differentiating layers of the wart. In HPV1 warts, the E4 protein is progressively cleaved and phosphorylated during terminal differentiation [6,25,37]. Thus, the full-length E4 molecule is present in the deepest wart layers where viral DNA replication is occurring whilst the more processed E4 polypeptides accumulate in cells where progeny virions are being assembled. A similar picture of regulated modification of E4 probably occurs in HPV16-induced lesions [26].

Interestingly, protein sequences that mediate biological activities and subcellular localization correspond to those regions that are modified through posttranslational mechanisms, i.e. self-multimerization, phosphorylation and proteolysis (Fig. 1). Taken together with the observation that modification of the E4 protein is influenced by epithelial differentiation, it is probable that E4 function is modified by post-translational events during the virus life-cycle [25,37,68].

E4 and the keratin cytoskeleton

The finding that HPV16 E4 associated with the keratin IFs and induced their disruption was an interesting development in E4 research, as it identified a host cell structure targeted by E4 [24]. Subsequent studies showed that other E4 proteins (Table 1), including those of cutaneous virus types, were also IF-associated proteins (IF-APs) [5,66,69]. Mutational analyses indicated that the leucine-rich cluster (LLXLL) located towards the N-terminus of many E4 proteins, including those identified as IF-APs, was required for E4's association with the keratin cytoskeleton, as mutants that had this motif deleted or mutated failed to align with the cytoskeleton [68]. It should be noted that not all E4 proteins contain a similar motif, e.g. HPV types 5 and 8, and whether these proteins associate with the keratin IFs remains to be seen. However, whilst the leucine motif represents an important structural motif in the E4-keratin IF association very little else is known about the mechanism of E4's interaction with the cytoskeleton, least of all the identity of E4-binding partners. A study of E4 isolated from HPV1 warts does not show the keratin proteins themselves as major E4 interacting factors [27], and neither did HPV1 E4 interact with pre-formed IFs *in vitro* [27], indicating that association with the IFs is likely to involve accessory proteins. One such candidate, filaggrin, a crosslinker of keratin fibres, binds to HPV1 E4, but only following incubation of the wart extracts at 4°C [27]. Another potential candidate, a cellular protein of unknown identity, interacts with HPV16 E4 in a yeast two-hybrid screen and antibodies specific for the cellular factor produced a cytoplasmic filamentous staining pattern in epithelial cells [23].

Based on current knowledge of E4 modification *in vivo*, the limited proteolysis of the N-terminus of E4 would remove the critical leucine motif [68] (Fig. 1), indicating that the interaction with the keratin cytoskeleton is probably specific for the full-length molecule only [68,70]. Thus, as expression of full-length protein is restricted to the lower suprabasal layers of warts, and processed species appear in upper layers, the E4-keratin cytoskeleton association may only be important for E4 function associated with early stages of the viral productive cycle [68].

An intriguing aspect of E4's function as an IF-AP is its induced reorganization of the keratin networks from a fine fibrous meshwork that runs throughout the cytoplasm into a juxtanuclear fibrous inclusion body [5,24,66,69] (Table 1). Whilst an N-terminal leucine-rich sequence is necessary for alignment between E4 and the IFs [68], sequences at the C-terminus are essential for induction of IF disruption. Fine mutational mapping of HPV16 E4 identified seven amino acids corresponding to the C-terminal tail that were involved in mediating the collapse of the cytoskeleton [70]. In fact, single substitutions of hydrophobic residues within this short stretch of amino acids was sufficient to abrogate E4's disruptive function despite alignment of the mutant protein along keratin fibres [70]. This region of HPV16 E4 is moderately conserved between E4 proteins of virus types that have a similar specificity for mucosal epithelia, and has also been implicated in self-multimerization of both HPV 11 and HPV16 proteins [11,70]. This suggests that an oligomeric form of E4 might mediate collapse of the keratin cytoskeleton. Not all E4 IF-APs however, induce collapse of the keratin cytoskeleton (Table 1). The HPV1 E4-keratin networks remain intact [69], and even when this protein has formed "*in vivo*-like" inclusions the networks are only partially disorganised [74]. Induction of IF collapse is not strictly type-specific function, as the E4 protein of the cutaneous HPV2 virus induces IF reorganization in a manner similar to the HPV16 protein [5]. Presumably, the fact that HPV2 E4 has greater homology with E4 proteins of anogenital types, including conservation of C-terminal sequences, accounts for HPV2 E4's behaviour [70]. The importance of these C-terminal sequences in mediating IF collapse is reinforced by the fact that HPV1 E4 does not have this homology and does not mediate IF collapse. Therefore, the observation that an HPV1 E4 mutant containing a deletion of a C-terminal motif (DL[D/E]X[Y/F]]), conserved between cutaneous types, induces keratin IF collapse in a manner remarkably similar to HPV16 E4 is intriguing [68]. An explanation of this phenomenon could be that loss of this region may induce conformational changes in the HPV1 molecule and/or modify interactions with cellular factors such that the mutant protein is able to function as a disrupter of the IF networks. If so, then it is tempting to speculate that perhaps the function of this region of E4 might be modified *in vivo* by posttranslational events.

It has been postulated that E4 acts to disturb the architecture of the keratin cytoskeleton in HPV productively-infected cells in order to render the cell's integrity unstable and prone to lysis, thereby aiding the release of the newly produced virions [24]. This is an attractive hypothesis in light of advances in understanding keratin function in epithelia—through the study of the many different forms of epidermolysis bullosa and epidermolytic hyperkeratosis (see the reviews of McLean and Lane, 1995 [55] and Fuchs, 1997 [36], and references therein). Clinical features of these skin diseases are typified by mechanical stress-induced skin blistering due to cytolysis of the keratinocytes, and electron microscopical examination of these cells shows clumps or aggregates of keratins. Studies have shown that individuals affected by these skin diseases carry mutations within their keratin genes. The mutant keratin proteins are ineffective in forming a cytoskeleton that is able to withstand physical trauma and consequently they breakdown. The E4-associated keratin aggregates formed in epithelial cells [24,66,69] are reminiscent of the keratin clumps observed in the epidermal cells of these skin disorders. However, whether E4 disrupts keratin IFs in the differentiating cells of the wart or tumour *in vivo* is not clear. In cutaneous warts, E4-associated inclusion bodies may be aggregates of E4 and keratin filaments; they are often closely associated with keratins bundles [16,74] and, in some lesions, the inclusions themselves have a filamentous appearance [16]. However, the true nature of the inclusions formed *in vivo* is not known and, as mentioned above, their formation in cultured expression systems is not associated with a dramatic rearrangement of IF networks [69,74]. It has been shown that inclusions formed in epithelial cells by an HPV1 E4 molecule lacking extreme N-terminal sequences (corresponding to the E1 derived sequences) are filamentous, and the filaments appear contiguous with keratin-like fibres [74]. Furthermore, juxtanuclear E4 filamentous structures were identified in parabasal cells of an HPV1-infected xenograft of human skin [69]. This observation is consistent with the notion that the E4–IF interaction is transient and may occur only in the initial stages of the productive phase of the virus. E4 proteins of mucosa-specific virus types, in general, have not been shown to localize to cytoplasmic inclusions. However, HPV16 E4 is associated with filaments, thought to be keratin IFs, in suprabasal keratinocytes [26,79]. These E4-labelled filaments are aggregated into bundles at the periphery of cells in the upper epithelial layers of tumours (our unpublished data). It would seem therefore, that although there are differences between E4's subcellular localization *in vivo* and in cultured monolayer cells, there is a close affinity between E4 (at least for some HPV types) and cytoskeletal structures in differentiating keratinocytes. However, it is not possible, at this stage of our understanding of E4 biology, to confirm or reject the hypothesis that E4 induces rearrangement of the keratin cytoskeleton *in vivo*. It is important to note that cultured monolayer epithelial cell systems do not recapitulate the phenotype of differentiating keratinocytes [35]—the natural site of E4 expression. For instance, different keratin subsets are expressed, and simple glandular keratins (8 and 18) are primarily expressed in cultured epithelial cells. These appear to be more sensitive to E4-induced collapse than other types of keratin networks [69]. In addition, host cell factors that are targeted by E4 may have variable intracellular locations between monolayer and differentiating keratinocytes.

The mode of E4 expression in experimental cell culture systems, as well as cell type, can influence both its subcellular localization and its effect on the keratin cytoskeleton. For instance, HPV1 E4 localizes to IFs when introduced into cells using recombinant SV40 [68,69] or vaccinia [27] viruses, but assembles into *in-vivo*-like inclusions following transient transfection using expression plasmids [73,74], possibly reflecting different levels of E4 protein production between the different systems. Furthermore, HPV31b E4 readily induces IF collapse when transiently expressed in SV40-immortalised cells, whilst no alignment with IFs was apparent when this protein was spontaneously induced in monolayer cultures of CIN-612 cells that contain the HPV31b genome [66].

A clearer understanding of the mechanism of E4's induced disruption of the keratins is necessary to appreciate whether this activity is an anomaly of E4 over-expression in epithelial cells grown in tissue culture or whether this is an important function for E4 in the virus life-cycle. Several reports show that other viruses encode proteins that disrupt IFs in epithelial cells. For example, the adenovirus L3 protease induces keratin network collapse by cleaving the keratin proteins [14], and the human immunodeficiency virus Vif protein induces aggregation of vimentin IFs [41]. But, as is the case for HPV, it is not clear why, or if, these viruses need to reorganize the IF cytoskeleton in the host cell. The IF cytoskeleton is a dynamic structure, and partial disruption of the networks is a normal physiological process that occurs during mitosis and in response to stress. Molecular chaperones are emerging as an important class of cellular proteins that facilitate formation of the IF cytoskeleton, and also act as protectors of these structures (reviewed by Liang and MacRae, 1997). It is therefore not inconceivable that E4 could disrupt IFs through interfering with the action of IF-specific chaperones.

Although maintaining a cell's structural integrity is a major function of the keratin IFs, they have been shown to modulate cell proliferation [64], interact with regulators of signal transduction pathways [32] and proteasomes involved in ubiquitin-mediated protein degradation [63]. Thus, E4 might interfere with these putative additional functions of the cytoskeleton. The keratin networks may act as a cytoplasmic anchor of E4, as loss of the leucine-rich motif causes the HPV1 and HPV16 proteins to accumulate in the nucleus [68,70]. HPV1 E4 structures, that could represent an early stage of inclusion assembly, have been observed to form along the length of the IF fibres [68,69]. This may indicate that the IF regulates formation of E4 oligomers–keratins have been shown to inhibit the formation of high molecular weight HPV1 E4 complexes *in vitro* [27]. Alternatively, but not mutually exclusive to this idea, the cytoskeleton could act as a scaffold to which E4 can attach host cell proteins, allowing their function to be modified or abrogated, perhaps by altering posttranslational modification of the protein or preventing them trans-locating to appropriate subcellular compartments. The later point is supported by a recent report of cyclin B/cdc2 co-localization to HPV16 E4/keratin networks in epithelial cells [19, permission to cite this reference given by the authors, and discussed later in the chapter]. Cyclin B/cdc2 is an important regulator of progression of the cell cycle into mitosis and perhaps E4 could interfere with the function of this complex by preventing its nuclear localization [19].

The physiological function of E4's association with keratin cytoskeletal elements remains to be determined. It would be surprising, in light of current data, if these cellular structures are not important in E4 function.

E4 and epithelial keratinization

Perturbation of normal epithelial differentiation characteristics is apparent in E4-positive cells, including a decrease or even loss of detectable differentiation-specific keratins, and structural proteins that are constituents of the cell envelope such as filaggrin, involucrin and loricrin [6,9,26,74,86]. This prompted the suggestion that E4 may interfere with aspects of the normal process of keratinization, altering the phenotype of the infected keratinocyte to one that supports virus replication [6] and virion maturation [21]. The keratinization process is necessary for the formation of a tough and durable epidermis (see the reviews of Watt, 1989 [88] and Fuchs, 1990 [35] for more details). The end result of this process is the formation of the cornified cell envelope (CCE), an insoluble, highly resistant structure, beneath the plasma membrane of the corneocyte. It consists of various cross-linked precursor proteins, including loricrin and involucrin. The filaggrin crosslinked keratin macrofibrils fill the CCE's to finally form enucleated squame [67,77,78]. A recent study has shown that the cornified squames isolated from HPV11-infected lesions are morphologically abnormal and more fragile compared to those from uninfected tissue [9], and can be used to infect a xenograft system with a high level of efficiency [10]. The HPV11 E4 protein was found in these aberrant cornified cells, and was associated with fragments of the CCE [9]. By crosslinking to constituents of these structures, E4 could disrupt their normal assembly [78] rendering them fragile and able to release progeny virions easily [9]. Disrupting the keratin fibres would also result in a more fragile cell, as discussed above, and it is possible that E4 acts at several points in the keratinization process, possibly mediated by different forms of E4 that arise through posttranslational modification.

Terminal differentiation of epidermal keratinocytes as cells move from the granular layers to the cornifying layers involves drastic morphological changes, regarded as a specialised form of cell death [39, 85]. Such a cellular environment, with significant protease and endonuclease activity, would not be conducive to virion synthesis, and it has been suggested that E4 interacts with these so-called death proteins, to inhibit or delay the onset of terminal differentiation [73]. Alternatively, or in addition, E4 could target proteins that have a role in initiating this event. Several studies suggest that filaggrin, and fragments of its precursor, profilaggrin, may have a role in facilitating the apoptotic process in keratinocytes [18,43,48]. Of interest then is the finding that HPV1 E4, when purified from warts co-elutes with filaggrin [27], and by interacting with filaggrin, E4 could prevent the infected cells becoming anucleated, and retention of the nuclei (parakeratosis) is a feature of HPV warts [73]. However, as mentioned previously the physiological significance of this association is uncertain [27], and furthermore the fact that filaggrin is not detectable in E4-positive cells of warts is inconsistent with an E4:filaggrin association [27,74].

It has been reported that the E6 oncoprotein can interfere with the normal differentiation programme [54,76,83], and specifically may inactivate the onset of programmed cell death [1]. Although these studies indicate that E6 can interfere with the differentiation programme they do not rule out a role for E4 in aspects of the alterations in keratinization observed in warts. E6 expression did not prevent up-regulation of differentiation specific marker proteins [1], and it is these proteins that are down-regulated in E4-positive cells in warts and tumours.

E4 and viral genome amplification

In an early study of E4 expression in HPV1 warts it was noted that there was a striking correlation between initiation of E4 protein expression and the onset of viral DNA amplification in suprabasal epithelial cells [6]. It was suggested that this could indicate a role for E4 in HPV genome amplification, perhaps interfering with viral or cellular mechanisms that inhibit genome amplification in the basal cells [6]. More recent studies have shown that a close relationship between the onset of these two viral events is common to other PV infections, including cutaneous warts induced by HPV types other than HPV1 [26], mucosal tumours induced by HPV16 [26], and some animal papillomavirus lesions [2,58]. Furthermore, several experimental PV genome replication systems have also shown that viral DNA amplification and E4 expression occur in the same cells [45,75], reinforcing the link between E4 protein and virus amplification. It has been reported that disruption of the major splice acceptor in the E4 ORF of the HPV31 genome abrogated the stable maintenance of episomal genomes in primary keratinocytes [47]. The virus needs to maintain its genome as an extrachromosomal plasmid in order to replicate and activate late gene expression [34]. Although, it was concluded that loss of episomal maintenance was most likely due to qualitative and/or quantitative effects on viral transcription [47], the possibility that loss of expression of a *bona fide* E4 molecule could also have contributed to the phenotype of these mutant genomes was not entirely disproven.

Of course, the idea that E4 plays a direct role in the replication process is not particularly attractive because PV DNA can be replicated efficiently in transient replication systems without the need of E4 protein, and neither is E4 necessary for establishment of BPV1 genomes in mouse cells [57]. However, there is evidence that these viruses switch to a different mode of DNA replication when amplifying their genomes in differentiating cells [31]. Therefore, the possibility that E4 could function directly in the replication process in naturally-occurring infections should not be overlooked at this stage of our understanding of E4 biology. In fact, nuclear forms of the protein are found in epithelial cells actively undergoing viral DNA amplification *in vivo* [6,21,58,61]. It is conceivable that the function of nuclear E4 could be to direct cellular (or viral) factors necessary for genome replication to sites of virus DNA synthesis. It is therefore of interest that the cellular replication factor, minichromosome maintenance protein 7 (Mcm7) interacts with HPV1 E4 through C-terminal residues (Fig. 1) in a yeast two-hybrid assay [3] and in cultured keratinocytes [72]. Mcm7 is part of a six-member MCM complex of proteins that is a component of pre-replication complexes (pre-RCs). These complexes assemble at cellular DNA origins in a regulated manner and "licenses" the origins for initiation of replication (replication licensing of chromosomal origins is reviewed by Lei and Tye, 2001 [50]). Whether pre-RC components are involved in replication of the PV genome is not known, but they are thought to be necessary for licensing and initiation of replication at *oriP* of Epstein–Barr virus (EBV), and there is evidence to suggest that the EBV protein EBNA1 might be responsible for recruiting these factors to the functional replicator [13]. The interaction between E4 and Mcm7 could be linked to E4's inhibitory activity on cellular DNA replication, and this is discussed later in this section. Interestingly, Mcm7 is degraded by HPV18 E6 via ubiquitination [49]; this lends support to the possibility that this cellular replication protein has some role in the PV life-cycle.

An indirect role for E4 in PV genome amplification by interacting with the keratinocyte cell and modifying its metabolism to one that supports PV replication, is possible. In an early study by Burnett and colleagues [45], it was noted that when E4 expression and viral DNA amplification were induced in BPV-1 transformed mouse C127 cells, E4 was most often expressed in cells in which E2 had accumulated into dense nuclear aggregates. In E4-negative cells however, E2 was localized to multiple small nuclear foci [45]. More recent studies have shown that BPV1 E2 accumulates into nuclear structures known as promyelocytic leukaemia (PML) oncogenic domains (PODs) (also known as NDIO bodies) through the action of L2 [20]. Furthermore, HPV11 E1 and E2 proteins, plus a replicating HPV11 origin-containing plasmid have been shown to be closely associated with these structures in epithelial cells [84]. It is conceivable therefore that the E2-stained nuclear foci observed in BPV1 transformed C127 cells were in fact E2-positive PODs. While the precise roles of PODs remain somewhat cryptic, there is evidence to suggest that they exert antiviral effects that must be overcome early in virus infection to establish productive replication (reviewed by Maul, 2000 [53]). A growing number of reports indicate that RNA and DNA viruses perhaps utilise PODs as sites for virus replication and the viruses encode proteins that can disrupt and reorganize these structures [44,52]. It has been suggested that PVs may also replicate (and assembly progeny) at, or in close proximity to these nuclear domains [20,59,84]. Thus, the interesting question is whether E4 functions to facilitate PV replication (and/or virus assembly) at PODs by restructuring these nuclear domains? The observation that E2-positive nuclear foci (PODs?) are reorganized in E4-positive C127 cells is consistent with this putative role of E4 [45]. Furthermore, localization of the POD-associated protein PML is altered in E4-positive cells of HPV1-induced warts and accumulates at the site of nuclear E4 inclusions (our unpublished data).

PVs are strictly dependent on host cell replication factors for replication of their genomes. In normal terminal differentiating epithelia these cellular factors are not present in the suprabasal keratinocytes—the site at which PV genome amplification occurs. Therefore, early virus proteins E6 and E7 interact with host cell cycle regulators in order to stimulate these cells to re-enter an S phase-like state and activate the cell's replication machinery (the reader should refer to Chapters 2, 4 and 5 for more details). Two recent reports show that E4 proteins may also interfere with normal cell cycle progression and the effect may facilitate a cellular environment conducive to virus replication. In the first report, HPV1 E4 was shown to significantly inhibit the G1 to S phase progression of synchronised fibroblasts that had been released from density dependent growth arrest in G0 [72]. Inhibition was dependent on a full-length E4 molecule as a deletion mutant protein that had lost the extreme N-terminal sequences (resembling the processed 16K species) proved to be less effective at preventing S phase transition. Addition of purified HPV1 E4 proteins into an in vitro mammalian cellular DNA replication system [80] demonstrated that the full-length E4 molecule (17K), but not the 16K species, specifically inhibits initiation of cellular DNA replication by nearly 90% [72]. Although the precise mechanistic details of E4's inhibitory activity are not yet known the interaction between HPV1 E4 and the pre-RC factor Mcm7 suggests that by physically interacting with pre-RC components, E4 could interfere with assembly or activation of the pre-RC and thereby inhibit cellular DNA replication [72]. The normal controls of DNA replication, including the "once-per-cell-cycle" licensing of DNA replication, are most likely disrupted in HPV-infected surprabasal keratinocytes, leading to unscheduled replication of the host genome [34,46,90]; this would deplete host cell factors essential for viral DNA synthesis. HPV1 E4, by inhibiting cellular DNA synthesis in these cells, would prevent depletion of host cell replication proteins and allow them to be recruited to sites of viral DNA synthesis. Such an hypothesis is not without precedent, for the immediate early 2 protein of the human cytomegalovirus induces cells to enter an S phase-like state with expression of cellular replication factors, but prohibits competitive cellular DNA synthesis [89]. In further support of the hypothesis, *in vivo* expression of the full-length inhibitory (17K) E4 molecule is restricted to cells amplifying the viral genome (our unpublished data). The extreme N-terminus that is required for effective inhibition is also necessary for association with keratin IF [68] (Fig. 1), but whether these activities are connected is not known at the present time.

The second report demonstrated that HPV16 E4 induces the fission yeast, *Sacchromyces pombe,* to arrest in the G2 phase of the cell cycle [19]. A G2 arrest was also observed following expression of the HPV16 protein in G1 synchronised human keratinocytes, a function that maps to a proline-rich region near to the N-terminus of the molecule (Fig. 1). The precise mechanistic details of the E4-induced arrest are not known, but as mentioned earlier in this chapter, the cell cycle proteins cyclin B and p34^{CDC2}, required for entry and progression through mitosis, are localized with HPV16 E4 to IFs [19]. Thus, failure of cells to progress through into mitosis might be due to E4-induced cytoplasmic retention of this cyclin complex [19]. This action of E4 could inhibit suprabasal keratinocytes committed to virus replication from moving through the cell cycle and entering mitosis.

These two recent studies represent interesting developments in understanding E4 functions and their outcome will no doubt provide an insight into the role this protein plays in the productive life-cycle. It is intriguing that the E4 protein of another anogenital virus type, HPV11, also induced a G2 arrest, but the cutaneous HPV1 E4 failed to do so [19]. Also, it is not known if other E4s, besides HPV1 E4, inhibit initiation of cellular DNA replication [72]. It could be that these are type-specific E4 functions that reflect differences in the site of virus genome amplification in the lesion. In cutaneous warts, E4 protein induction and viral DNA amplification occurs in cells immediately above the proliferating basal layers—the parabasal cells. These cells have only just started to differentiate. In contrast, in lesions induced by mucosa-specific viruses these events occur much higher in the differentiating layers—the cells of the upper spinous layers and granular layers [6,26].

E4 and capsid expression, and virion assembly

In natural PV infections, virion assembly occurs in E4-positive cells, and thus E4 could be necessary for aspects of this late viral process. Interestingly, expression of E4 and L1 is linked as the major capsid protein L1 is translated from a bicistronic doubly spliced message, that also contains the E1^E4 gene, E1^E4^L1. Although it is not clear why the major capsid protein is translated in a coordinated manner with E4, cell-free translation studies show that HPV11 E4 expression is greater from the doubly spliced transcript than a E1^E4 construct [8]. Also, the E1^E4 ORF it is translated preferentially to that coding for L1 [8]. Late PV gene expression in the differentiating keratinocyte is primarily regulated at the post-transcriptional level, involving mRNA splicing, message stability and degradation and RNA transport, and involves host cellular factors. For further details of the regulation and control of late PV gene expression the reader should refer to Chapter 3. It is possible that E4 influences expression of the PV late genes at a posttranscriptional level as it has recently been demonstrated that HPV16 E4 interacts with a novel putative DEAD box RNA helicase, referred to as E4-DBP [23]. DEAD box proteins are involved in a number of cellular processes including mRNA splicing, ribosomal assembly, initiation of translation and RNA transport. Although the precise function of E4-DBP was not established in this study it was shown to share homology with bacterial and yeast proteins (SrmB, DeaD, and RhlB) that are involved in the regulation of mRNA degradation and stability, and ribosome biogenesis [23]. Furthermore, E4-DBP binds to RNA including the major late transcript of HPV16, has ATPase activity that is partially inhibited by HPV16 E4, and it is expressed in E4-containing cells *in vivo* [23]. Other viruses can target similar cellular proteins, for instance, Hepatitis C virus core protein interacts with the DEAD box protein DDX3 in mammalian cells [60]. Thus, targeting members of this family of proteins may represent a general mechanism by which viruses can modulate expression of host and/or viral genes. It is important to note however that the E4 proteins of HPV1, nor the more closely related HPV6, did not interact with E4-DBP [23], suggesting that this is not a common activity amongst PV types, unless they interact with different members of the family of DEAD box proteins.

A direct link between E4 and the process of genome encapsidation and assembly of the virus coat is not obvious. *In vitro* studies of capsid assembly do not show a requirement for E4 in the assembly of the capsid proteins L1 and L2 into virus-like particles [91] however, whether this is also true *in vivo* is not known. Studies in rodent fibroblasts that maintain episomal copies of the BPV1 genome have shown that L2 and E2 are linked to encapsidation of the genome [20,59]. The sub-nuclear domains known as PODs may act as sites of virion assembly [20,59], as well as virus replication [84], as previously mentioned, and as suggested earlier, if E4 has a role in reorganization of these domains then this may influence assembly of the newly synthesized progeny.

Concluding remarks

E4 biological activities can be broadly divided into two groups. The first group comprises interactions between E4 and epithelial cell specific structures, such as keratin IFs and the CCE. E4 interactions and activities involving cellular mechanisms that are central to normal cell growth, such as chromosomal replication, mitosis, control of mRNA processing and protein translation, form the second group. This profile of E4 biological functions strongly supports the notion of a central role for E4 in modifying the host keratinocyte to a phenotype that enables the virus to maximise its own replication, facilitate production of progeny and promote their effective transmission. However, between individual PVs, this common role of E4 is most likely executed using non-identical, but overlapping, mechanisms and this probably reflects differences in the biology of these viruses. This notion is supported by the findings that some of E4's activities are virus type-specific. For example, the interaction with a putative RNA helicase (HPV16 E4-specific). Others, such as association with the keratin cytoskeleton, does not appear to be restricted to E4s of a particular PV sub-group, although how they interact with these cellular structures may not be identical between virus types. This variation in E4 function between virus types is not unexpected. All PVs have to successfully replicate their genome, express late genes and assemble progeny in the differentiating keratinocyte. To achieve this they uncouple cell cycle progression from differentiation. The E6 and E7 proteins play a central role in this activity, but just as these viral proteins of different PV groups have evolved to disrupt different cellular pathways to achieve a similar goal, E4 too has probably evolved to function by targeting different cellular processes.

The emerging profile of E4 activities reveals an interesting relationship between E4 structure and function. To date, the N-terminal domain in particular is essential for a wide range of E4 activities, including interaction with the keratin cytoskeleton and inhibition of chromosomal replication (Fig. 1). This region is significantly modified during the life-cycle of the virus, and it would be predicted that this would disrupt E4 functions that are mediated by this domain. Posttranslational modification of E4 appears to occur in a regulated manner in the PV-induced lesion related to stages of the virus life-cycle and epithelial differentiation. E4 therefore probably performs its role through a series of modified polypeptides that encode different biological activities. If this is true then it is important that we establish whether the processed E4 polypeptides encode different biological activities. Technologies such as laser capture microscopy, used in conjunction with proteomics, will help to characterize the relationship between E4 function and protein modification *in vivo*. As well, elucidation of the three-dimensional structure of the E4 proteins will enable us to establish the molecular basis for their mode of action.

Our knowledge of E4 biological activities has been compiled from studies primarily using epithelial cell culture systems in which E4 is expressed in isolation of other PV proteins, and perhaps just as significant, in undifferentiated cells. Whilst these have been enormously informative, the natural functions of E4 might be partially (or completely) obscured, in these systems. Therefore, it is important that use is now made of epithelial cells containing episomal copies of PV genomes, which respond to *in vitro* differentiation signals, as well as some of the animal models of PV infection, to explore E4 function in the context of a productive infection. The importance of these systems is highlighted by a recent study that reports that cottontail rabbit papillomas produced by a mutant cottontail rabbit PV genome containing a knockout of the E4 gene did not show evidence of viral genome amplification or expression of capsid proteins ([65], cited with permission from the authors). These findings strongly support the supposition that E4 functions are necessary for the productive phase of the PV life-cycle.

Inasmuch as E4 has not attracted much attention, because it has no apparent role in oncogenicity, the fact that its expression is not retained in cancers could be linked to its effects on cell growth. Inhibition of both chromosomal replication and mitotic division would be considered refractory to malignant progression. In the future, the development of mimetic compounds based on E4's mode of action to inhibit cellular DNA synthesis could be used in the treatment of hyperproliferative diseases such as cancer.

Acknowledgements

The author would like to thank Roger Grand and Phillip Gallimore for critical reading of the manuscript.

References

1. Alfandari J, Magal SS, Jackman A, Schlegel R, Gonen P, Sherman L. HPV16 E6 oncoprotein inhibits apoptosis induced during serum-calcium differentiation of foreskin keratinocytes. Virology 1999; 257: 383–396.
2. Anderson RA, Scobie L, O'Neil BW, Grindlay GJ, Campo MS. Viral proteins of bovine papillomavirus type 4 during the development of alimentary canal tumours. Vet J 1997; 154: 69–78.
3. Ashmole I. Ph.D. thesis, 1997, University of Birmingham.
4. Ashmole I, Gallimore PH, Roberts S. Identification of conserved hydrophobic C-terminal residues of the human papillomavirus type 1 E1 ^ E4 protein necessary for oligomerization *in vivo*. Virology 1998; 240: 221–231.
5. Berthoud T, Roberts, S. Unpublished data.
6. Breitburd F, Croissant O, Orth G. Expression of human papillomavirus type-1 E4 gene products in warts. Cancer Cells (Cold Spring Harbor) 1987; 5: 115–122.
7. Brown DR, Fan L, Jones J, Bryan J. Colocalization of human papillomavirus type 11 E1 ^ E4 and L1 proteins in human foreskin implants grown in athymic mice. Virology 1994; 201: 46–54.
8. Brown DR, Pratt L, Bryan JT, Fife KH, Jansen K. Virus-like particles and E1 ^ E4 protein expressed from the human papillomavirus type 11 bicistronic E1 ^ E4 ^ L1 transcript. Virology 1996; 222: 43–50.
9. Bryan JT, Brown. DR. Association of the human papillomavirus type 11 E1 ^ E4 protein

with cornified cell envelopes derived from infected genital epithelium. Virology 2000; 277: 262–269.

10. Bryan JT, Brown DR. Transmission of human papillomavirus type 11 infection by desquamated cornified cells. Virology 2001; 281: 35–42.

11. Bryan JT, Fife KH, Brown DR. The intracellular expression pattern of the human papillomavirus type 11 E1^E4 protein correlates with its ability to self associate. Virology 1998; 241: 49–60.

12. Bryan JT, Han A, Fife KH, Brown. D. 2000; The human papillomavirus type 1 E1^E4 protein is phosphorylated in genital epithelium. 268: 430–439.

13. Chaudhuri B, Xu H, Todorov I, Dutta A, Yates JL. Human DNA replication initiation factors, ORC and MCM, associate with *oriP* of Epstein–Barr virus. Proc Natl Acad Sci USA 2001; 98: 10085–10089.

14. Chen PH, Ornelles DA, Shenk T. The adenovirus L3 23 kilodalton proteinase cleaves the amino-terminal head domain from cytokeratin 18 and disrupts the cytokeratin network of HeLa cells. J Virol 1993; 67: 3507–3514.

15. Chow LT, Reilly SS, Broker TR, Taichman LB. Identification and mapping of human papillomavirus type 1 RNA transcripts recovered from plantar warts and infected epithelial cell cultures. J Virol 1987; 61: 1913–1918.

16. Croissant O, Breitburd F, Orth G. Specificity of cytopathic effect of cutaneous human papillomaviruses. Clinics in Dermatol 1985; 3: 43–55.

17. Crum CP, Barber S, Symbula M, Snyder K, Saleh AM, Roche JK. Coexpression of the human papillomavirus type 16 E4 and L1 open reading frames in early cervical neoplasia. Virology 1990; 178: 238–246.

18. Dale BA, Presland RB, Lewis SP, Underwood RA, Fleckman P. Transient expression of epidermal filaggrin in cultured cells causes collapse of intermediate filament networks with alteration of cell shape and nuclear integrity. J Invest Dermatol 1997; 108: 179–187.

19. Davy C, Raj K, Masterson P, Miller J, Jackson D, Zumbach K, Cuthill S, Doorbar J. The E1^E4 protein of HPV16 causes human keratinocytes and *S. pombe* to arrest at G2 with redistribution of cyclin B and p34^{CDC2} to the insoluble fraction. 2000 International Papillomavirus Conference, Barcelona, Spain.

20. Day PM, Roden RB, Lowy DR, Schiller JT. The papillomavirus minor capsid protein, L2, induces localization of the major capsid protein, L1, and the viral transcription/replication protein, E2, to PML oncogenic domains. J Virol 1998; 72: 142–50.

21. Doorbar J, Campbell D, Grand RJA, Gallimore PH. Identification of the human papillomavirus-1a E4 gene products. EMBO J 1986; 5: 355–365.

22. Doorbar J, Coneron I, Gallimore PH. Sequence divergence yet conserved physical characteristics among the E4 proteins of cutaneous human papillomaviruses. Virology 1989; 172: 51–62.

23. Doorbar J, Elston RC, Napthine S, Raj K, Medcalf E, Jackson D, Coleman N, Griffin HM, Masterson P, Stacey S, Mengistu Y, Dunlop J. The E1^E4 protein of human papillomavirus type 16 associates with a putative RNA helicase through sequences in its C terminus. J Virol 2000; 74: 10081–10095.

24. Doorbar J, Ely S, Sterling J, McLean C, Crawford L. Specific interaction between HPV-16 E1-E4 and cytokeratins results in collapse of the epithelial cell intermediate filament network. Nature 1991; 352: 824–827.

25. Doorbar J, Evans HS, Coneron I, Crawford LV, Gallimore PH. Analysis of HPV-1 E4

138

gene expression using epitope-defined antibodies. EMBO J 1988; 7: 825–833.

26. Doorbar J, Foo C, Coleman N, Medcalf L, Hartley O, Prospero T, Napthine S, Sterling J, Winter G, Griffin H. Characterization of events during the late stages of HPV16 infection *in vivo* using high affinity synthetic Fabs to E4. Virology 1997; 238: 40–52.

27. Doorbar J, Medcalf E, Napthine S. Analysis of HPV1 E4 complexes and their association with keratins *in vivo*. Virology 1996; 218: 114–126.

28. Doorbar J, Myers G. The E4 protein. In: G Myers, H Delius, J Icenogel, H-U Bernard, C Baker, A Halpern, C Wheeler (Ed), Human Papillomaviruses 1996, Vol III. Los Alamos National Laboratory, New Mexco, 1996, pp. 58–80.

29. Egawa K, Iftner A, Doorbar J, Honda Y, Iftner T. Synthesis of viral DNA and late capsid protein L1 in parabasal spinous cell layers of naturally occurring benign warts infected with human papillomavirus type 1. Virology 2000; 268: 281–293.

30. Eriksson A, Herron JR, Yamada T, Wheeler CM. Human papillomavirus type 16 variant lineages characterized by nucleotide sequence analysis of the E5 coding segment and the E2 hinge region. J. Gen. Virol 1999; 80: 595–600.

31. Flores ER, Lambert PF. Evidence for a switch in the mode of human papillomavirus type 16 DNA replication during the viral life cycle. J Virol 1997; 71: 7167–7179.

32. Foisner R. Dynamic organization of intermediate filaments and associated proteins during the cell cycle. Bioesssays 1997; 19: 297–304.

33. Frattini MG, Hurst SD, Lim HB, Swaminathan S, Laimins LA. Abrogation of a mitotic checkpoint by E2 proteins from oncogenic human papillomaviruses correlates with increased turnover of the p53 tumour suppressor protein. EMBO J 1997; 16: 318–331.

34. Frattini MG, Lim HB, Laimins LA. *In vitro* synthesis of oncogenic human papillomaviruses requires episomal genomes for differentiation-dependent late expression. Proc Natl Acad Sci USA 1996; 93: 3062–3067.

35. Fuchs E. 1990; Epidermal differentiation. Curr Opion Cell Biol 2: 1028–1035.

36. Fuchs E. Of mice and men: genetic disorders of the cytoskeleton. Mol Biol Cell 1997; 8: 189–203.

37. Grand RJA, Doorbar J, Smith KJ, Coneron I, Gallimore PH. Phosphorylation of the human papillomavirus type 1 E4 proteins *in vivo* and *in vitro*. Virology 1989; 170: 201–213.

38. Grassman K, Rapp B, Maschek H, Petry KU, Iftner T. Identification of a differentiation-inducible promoter in the E7 open-reading frame of human papillomavirus type 16 (HPV-16) in raft cultures of a new cell line containing high copy numbers of episomal HPV-16 DNA. J Virol 1996; 70: 2339–2349.

39. Haake AR, Polakowska RR. Cell death by apoptosis in epidermal biology. J Invest Dermatol 1993; 101: 107–112.

40. Heino P, Zhou J, Lambert PF. Interaction of the papillomavirus transcription/replication factor, E2, and the viral capsid protein, L2. Virology 2000; 276: 304–314.

41. Henzler T, Harmache A, Herrmann H, Spring H, Suzan M, Audoly G, Panek T, Bosch V. Fully functional, naturally occurring and C-terminally truncated variant human immunodeficiency virus (HIV) Vif does not bind to HIV Gag but influences intermediate filament structure. J Gen Virol 2001; 82: 561–573.

42. Hummel M, Hudson JB, Laimins LA. Differentiation-induced and constitutive transcription of human papillomavirus type31b in cell lines containing viral episomes. J Virol 1992; 66: 6070–6080.

43. Ishida-Yamamoto A, Takahashi H, Presland RB, Dale BA, Iizuka H. Translocation of profilaggrin N-terminal domain into keratinocyte nuclei with fragmented DNA in normal human skin and loricrin keratoderma. Lab Invest 1998; 78: 1245–1253.

44. Ishov AM, Maul GG. The periphery of nuclear domain 10 (ND10) as sites of DNA virus deposition. J Cell Biol 1996; 134: 815–826.

45. Jareborg N, Burnett S. Immunofluorescent detection of bovine papillomavirus E4 antigen in the cytoplasm of cells permissive *in vitro* for viral DNA amplification. J Gen Virol 1991; 72: 2269–2274.

46. Jian Y, Van Tine BA, Chien W-M, Shaw GM, Broker TR, Chow LT. Concordant induction of cyclin E4 and p21[cip1] in differentiated keratinocytes by the human papillomavirus E7 protein inhibits cellular and viral DNA synthesis. Cell Growth Diff 1999; 10: 101–111.

47. Klumpp DJ, Stubenrauch F, Laimins LA. Differential effects of the splice acceptor at nucleotide 3295 of human papillomavirus type 31 on stable and transient viral replication. J Virol 1997; 71: 8186–8194.

48. Kuechle MK, Presland RB, Lewis SP, Fleckman P, Dale BA. Inducible expression of filaggrin increases keratinocyte susceptibility to apoptotic cell death. Cell Death Diff 2000; 7: 566–573.

49. Kuhne C, Banks L. E3-ubiquitin ligase/E6-AP links multicopy maintenance protein 7 to the ubiquitination pathway by a novel motif, the LG2 box. J Biol Chem 1998; 273: 34302–34309.

50. Lei M, Tye BK. Initiating DNA synthesis: from recruiting to activating the MCM complex. J Cell Sci 2001; 114: 1447–1454.

51. Liang P, MacRae TH. Molecular chaperones and the cytoskeleton. J Cell Sci 1997; 110: 1431–1440.

52. Maul GG. Nuclear domain 10, the site of DNA virus transcription and replication. Bioessays 1998; 20: 660–667.

53. Maul G, Negorev GD, Bell P, Ishov AM. Properties and assembly mechanisms of ND10, PML bodies, or PODs. J Struct Biol 2000; 129: 278–287.

54. McCance DJ, Kopan R, Fuchs E, Laimins LA. Human papillomavirus type 16 alters human epithelial cell differentiation *in vitro*. Proc Natl Acad Sci USA 1988; 85: 7169–7173.

55. McLean WHI, Lane EB. Intermediate filaments in disease. Curr Opin Cell Biol 1995; 7: 118–125.

56. Nasseri M, Hirochika R, Broker TR, Chow LT. A human papillomavirus type 11 transcript encoding an E1 ^ E4 protein. Virology 1987; 159: 433–439.

57. Neary K, Horwitz BH, DiMaio D. Mutational analysis of open reading frame E4 of bovine papillomavirus type 1. J Virol 1987; 61: 1248–1252.

58. Nicholls PK, Doorbar J, Moore RA, Peh W, Anderson DM, Stanley MA. Detection of viral DNA and E4 proteins in basal keratinocytes of experimental canine oral papillomavirus lesions. Virology 2001; 284: 82–98.

59. Okun MM, Day PM, Greenstone HL, Booy FP, Lowy DR, Schiller JY, Roden RB. L1 interaction domains of papillomavirus L2 necessary for viral genome encapsidation. J Virol 2001; 75: 4332–4342.

60. Owsianka AM, Patel AH. Hepatitis C virus core protein interacts with a human DEAD box protein DDX3. Virology 1999; 257: 330–340.

61. Palefsky JM, Winkler B, Rabanus J-P, Clark C, Chan S, Nizet V, Schoolnik GK. Characterization of *in vivo* expression of the human papillomavirus type 16 E4 protein in cervical biopsy tissues. J Clin Invest 1991; 87: 2132–2141.

62. Palermo-Dilts DA, Broker TR, Chow LT. Human papillomavirus type 1 produces redundant as well as polycistronic mRNAs in plantar warts. J Virol 1990; 64: 3144–3149.

63. Palmer A, Mason GFG, Paramio JM, Knecht EW, Rivett AJ. Changes in proteasome localization during cell cycle. Eur J Cell Biol 1994; 64: 163–175.

64. Paramio JM, Casanova ML, Segrelles C, Mittnacht S, Lane EB, Jorcano JL. Modulation of cell proliferation by cytokeratins K10 and K16. Mol Cell Biol 1999; 19: 3086–3094.

65. Peh W, Brandsma J, Cladel N, Christensen N, Doorbar J. Production of papillomas in cottontail and domestic rabbits using CRPV knockout mutants reveals an essential role for E4 in viral replication and capsid synthesis. DNA tumour virus meeting, Cambridge, UK, 2001.

66. Pray TR, Laimins LA. Differentiation-dependent expression of E1 ^ E4 proteins in cell lines maintaining episomes of human papillomavirus type 31b. Virology 1995; 206: 679–685.

67. Reichert U, Michel S, Schmidt R. The cornified envelope: a key structure of terminally differentiating keratinocytes. In: M. Darmon and M. Blumenberg (Eds), Molecular Biology of the Skin. Academic Press, New York, 1993, pp. 107–150.

68. Roberts S, Ashmole I, Gibson LJ, Rookes SM, Barton GJ, Gallimore PH. Mutational analysis of human papillomavirus E4 proteins: Identification of structural features important in the formation of cytoplasmic E4/cytokeratin networks in epithelial cells. J Virol 1994; 68: 6432–6445.

69. Roberts S, Ashmole I, Johnson GD, Kreider JW, Gallimore PH. Cutaneous and mucosal papillomavirus E4 proteins form intermediate filament-like structures in epithelial cells. Virology 1993; 197: 176–187.

70. Roberts S, Ashmole I, Rookes SM, Gallimore PH. Mutational analysis of the human papillomavirus type 16 E1 ^ E4 protein shows that the C terminus is dispensable for keratin cytoskeleton association but is involved in inducing disruption of the keratin filaments. J Virol 1997; 71: 3554–3562.

71. Roberts S, Ashmole I, Sheehan TMT, Davies AH, Gallimore PH. Human papillomavirus type 1 E4 protein is a zinc-binding protein. Virology 1994; 202: 865–874.

72. Roberts S, Stoeber K, Gallimore PH, Williams GH. The HPV1 E1 ^ E4 protein inhibits initiation of cellular DNA replication. International Papillomavirus Conference, Barcelona, Spain, 2000.

73. Rogel-Gaillard C, Breitburd F, Orth G. Human papillomavirus type 1 E4 proteins differing by their N-terminal ends have distinct cellular localizations when transiently expressed *in vitro*. J Virol 1992, 66: 816–823.

74. Rogel-Gaillard C, Pehau-Arnaudet G, Breitburd F, Orth G. Cytopathic effect in human papillomavirus type 1-induced inclusion warts: *in vitro* analysis of the contribution of two forms of the viral E4 protein. J Invest Dermatol 1993; 101: 843–851.

75. Ruesch MN, Stubenrauch F, Laimins LA. Activation of papillomavirus late gene transcription and genome amplification upon differentiation in semisolid medium is coincident with expression of involucrin and transglutaminase but not keratin-10. J Virol

1998; 72: 5016–5024.

76. Sherman L, Schlegel R. Serum and calcium-induced differentiation of human keratino-cytes is inhibited by the E6 oncoprotein of human papillomavirus type 16. J Virol 1996; 70: 3269–3279.

77. Steinert PM, Marekov LN. The proteins elafin, filaggrin, keratin intermediate filaments, loricrin and SPRs are isodipeptide cross-linked components of the human epidermal cornified cell envelope. J Biol Chem 1995; 270: 17702–17711.

78. Steinert PM, Marekov LN. Initiation of assembly of the cell envelope barrier structure of stratified squamous epithelia. Mol Biol Cell 1999; 10: 4247–4261.

79. Sterling JC, Skepper JN, Stanley MA. Immunoelectron microscopical localization of human papillomavirus type 16 L1 and E4 proteins in cervical keratinocytes cultured in vivo. J Invest Dermatol 1993; 100: 154–158.

80. Stoeber K, Mills AD, Kubota Y, Krude T, Romanowski P, Marheineke K, Laskey RA, Williams GH. Cdc6 protein causes premature entry into S phase in a mammalian cell-free system. EMBO J 1998; 17: 7219–7229.

81. Stoler MH, Whitbeck A, Wolinsky SM, Broker TR, Chow LT, Howett MK, Kreider JW. Infectious cycle of human papillomavirus type 11 in human foreskin xenografts in nude mice. J Virol 1990; 64: 3310–3318.

82. Stoler MH, Wolinsky SM, Whitbeck A, Broker TR, Chow LT. Differentiation-linked human papillomavirus types 6 and 11 transcription in genital condylomata revealed by in situ hybridisation with message-specific RNA probes. Virology 1989; 172: 331–340.

83. Stöppler MC, Ching K, Stöppler H, Clamcy K, Schlegel R, Icenogle J. Natural variants of the human papillomavirus type 16 E6 protein differ in their abilities to alter keratinocyte differentiation and to induce p53 degradation. J Virol 1996; 70: 6987–6993.

84. Swindle CS, Zou N, Van Tine BA, Shaw GM, Engler JA, Chow LT. Human papilloma-virus DNA replication compartments in a transient DNA replication system. J Virol 1999; 73: 1001–1009.

85. Takahashi H, Aoki N, Nakamura S, Asano K, Ishida-Yamamoto A, Iizuka H. Cornified cell envelope formation is distinct from apoptosis in epidermal keratinocytes. J Dermatol 2000; 23: 161–169.

86. Tinsley JM, Fisher C, Searle PF. Abnormalities of epidermal differentiation associated with expression of human papillomavirus type 1 early region in transgenic mice. J Gen Virol 1992; 73: 1251–1260.

87. Tomita Y, Fuse A, Sekine H, Shirasawa H, Simizu B, Sugimoto M, Funahashi S. Human papillomavirus type 6 and 11 E4 gene products in condyloma acuminata. J Gen Virol 1991; 72: 731–744.

88. Watt FM. Terminal differentiation of epidermal keratinocytes. Curr Opion Cell Biol 1989; 1: 1107–1115.

89. Wiebusch L, Hagemeier C. The human cytomegalovirus immediate early 2 protein dissociates cellular DNA synthesis from cyclin-dependent kinase activation. EMBO J 2001; 20: 1086–1098.

90. Williams GH, Romanowski P, Morris L, Madine M, Mills AD, Stoeber K, Marr J, Laskey RA, Coleman N. Improved cervical smear assessment using antibodies against proteins that regulate DNA replication. Proc Natl Acad Sci USA 1998; 95: 14932–14937.

91. Zhou J, Sun XY, Stenzel D, Frazer IH. Expression of vaccinia recombinant HPV16 L1 and L2 ORF proteins in epithelial cells is sufficient for assembly of HPV virion-like particles. Virology 1991; 185: 251–257.

92. Zhu Q.-L, Smith TF, Lefkowitz EJ, Chow LT, Broker TR. Nucleic acid and protein sequence alignments of human and animal papillomaviruses constrained by functional sites. The University of Alabama at Birmingham Press, Birmingham, 1994.

an Papillomaviruses
McCance (editor)
02 Elsevier Science B.V. All rights reserved

e E5 protein of papillomaviruses

lo Venuti[1] and M. Saveria Campo[2]*

oratory of Virology, Regina Elena Cancer Institute, Viale Regina Elena 291, 00161 Rome, Italy
tute of Comparative Medicine, Department of Veterinary Pathology, Glasgow University, Garscube
e, Glasgow G61 1QH, Scotland, UK

oduction

lies of the mechanisms used by viruses to transform cells have generated much rmation on cell growth, gene expression, cell differentiation and cancer gression. DNA tumour viruses encode proteins that induce proliferation of the cells, and often their transformation, either by neutralizing cellular tumor pressor proteins such as p53 and p105Rb, or by activating cellular growth stimu-ry pathways. The papillomavirus (PV) E5 protein belongs to the class of viral eins which induce unscheduled cell proliferation. It is a protein with unusual acteristics displaying pleiotropic functions, the understanding of which is essary for a full knowledge of papillomavirus biology and the interactions reen virus and the host cell.

n this chapter we will analyse the biochemical and biological functions of E5 and highlight differences and similarities between E5 proteins from bovine llomaviruses (BPVs) and human papillomaviruses (HPVs).

E5 ORF

E5 open reading frame (ORF) commonly overlaps the E2 ORF in the early on of the PV genome. However, in the subgroup B epitheliotropic BPVs the E5 *(previous named E8) is located 5' to the E7 ORF and replaces the E6 ORF [54]. uence analysis suggested that the unusual location of the E5 ORF in these BPVs have been caused by intra- or inter-genomic recombination, leading to trans-tion of the E5 ORF and loss of the E6 ORF [54]. Interestingly, the cottontail it papillomavirus genome has two E5-like ORFs, one at the "classical" 3' end, one at the "non-canonical" 5' end of the early region [47], suggesting either gene ication or translocation without loss of the E6 ORF.

Despite the presence of the E5 ORF in the genome of BPVs and genital HPVs, E5 ORF is absent in the genome of other PVs, indicating that the protein is not ntial for either the life cycle of, or cell transformation by, these viruses.

The E5 protein

The E5 proteins range in size from approximately 40 amino acids in BPVs (42 for BPV-4 and 44 for BPV-1) to approximately 80 amino acids in HPVs (a noticeable exception is the E5 protein of HPV-83 which is only 47 amino acids long [14]). Despite few sequence similarities among E5 proteins from different PVs, E5 proteins are believed to have a common structure as they are all hydrophobic with a hydrophilic C-terminus domain. E5 is postulated to assume an α-helical configuration with one *trans*-membrane span for BPV E5 (Fig. 1) [104] and three for HPV E5 [15,111]. Immunoelectron and immunofluorescence microscopy studies have shown that, in agreement with their hydrophobicity and postulated structure, E5 proteins are localised in the endomembrane compartments (Golgi apparatus and endoplasmic reticulum) and occasionally, when overexpressed in cell cultures, in the plasma membrane [16,79]. Computational and NMR analyses of BPV-1 E5 have demonstrated that E5 is a type II transmembrane protein which forms dimers as paired left-handed α-helices with the carboxyl termini facing the lumen [104]. A genetic approach utilising a heterologous dimerization domain to force E5 monomers to adopt various orientations relative to one another within the dimer [67] confirmed the orientation described above as this is the only one that displays significant transforming activity (see below).

The E5 protein of BPV-1 migrates as an approximately 7 kDa monomer in reducing conditions and as a dimer of 14 kDa in neutral condition [16,95]. No

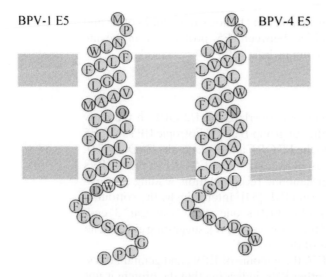

Fig. 1. Secondary structure of the BPV-1 and BPV-4 E5 proteins [54,94]. Residues in green are crucial for the functioning of the proteins. In BPV-1 E5 glutamine 17 mediates binding to both PDGF-R and 16kDa

evidence for post-translational modification of E5 protein, e.g. phosphorylation, has been reported [39]. The hydrophilic carboxyl-terminus includes two cysteine residues, C37 and C39, that stabilize the formation of homodimer structures via disulfide bonds, hence the presence of 14 kDa structures. Mutations of either cysteine reduce dimerization while double mutations abolish dimerization altogether [43]. HPV E5 migrates as a monomeric form of approximately 10 kDa (HPV-16) or approximately 12 kDa (HPV-6). There is one report of HPV E5 dimers observed in reducing gels of proteins from HPV-16 positive cervical scrapes, indicating a dimerization mechanism other than disulfide bonding [56]. However, *in vitro* expressed HPV-16 E5, either wild-type or epitope-tagged, has been found only as a monomeric band [26,53].

Biological characteristics of E5

In naturally infected tissues, E5 is expressed at low levels in the undifferentiated basal cells of the lower third of the epithelium [3,17,18]. However, BPV-1 E5 has been observed in higher amounts in the differentiated keratinocytes of skin warts, where it shows a granular staining pattern associated with the sites of viral capsid synthesis [17]. Similarly, HPV-16 E5 has been detected occasionally throughout the whole epithelium in high-grade cervical intraepithelial neoplasia and in invasive cancers still maintaining episomal HPV genomes [18,51,52]. Notably, E5 expression appears to correlated with that of type I growth factor receptors, i.e. epidermal growth factor receptor (EGF-R) [18], in apparent agreement with the functions of E5 as established *in vitro* (see below).

While the PV oncoproteins E6 and E7 are present throughout the course of the disease and their functions are necessary for the maintenance of a transformed state, expression of E5 takes place early in infection, and is often, but not always, extinguished in frank cancers. In human genital lesions, the expression of E5 is extinguished as the lesion progresses to malignancy, due to the frequent occurrence of the integration of the viral genome into the host chromosome at the E2/E5 ORF; in the lesions induced by BPV the mechanisms that silence the expression of E5 from an intact episomal genome are not known.

In partial agreement with what is observed in naturally occurring lesions, the E5 protein of HPV-31 was found present throughout the virus life cycle in organotypic raft cultures, with the highest level coincidental with the production of virus capsid [68].

Although E5 functions have not been established unequivocally *in vivo*, the summation of the results described above suggests that the presence of E5 in the basal layers contribute to the sustained proliferative state of the undifferentiated keratinocytes, while expression in differentiated keratinocytes would play a role in virus maturation, such as the amplification of viral DNA and/or anti-apoptotic action during vegetative replication (see later). However, further studies are needed to clarify this issue.

In vitro cell transformation

Despite the lack of amino acid sequence conservation among the various E5 proteins, several functional similarities have been observed in studies of cell transformation *in vitro*, where E5's role is becoming more defined.

For the past 20 years, BPV-1 has been a useful model for studying PV-induced cell transformation because of its capacity to induce transformation of murine fibroblasts in cell culture. By this assay E5 was recognised as the major transforming protein of BPV-1. BPV-1 E5 induces morphologic and tumorigenic transformation of rodent and human fibroblasts [94,95]. It increases focus formation in NIH 3T3 cells when co-expressed with receptor protein tyrosine kinases (PTK receptors) [66] and promotes DNA synthesis in quiescent fibroblasts [96]. BPV-4 E5 cooperates with activated *ras* in transforming primary bovine fibroblasts where it induces anchorage independence [55,79], and also transforms NIH 3T3 fibroblasts without the need for other oncogenes [75].

The E5 proteins encoded by HPVs display weak transforming activity. Experiments with HPV-6 provided the first evidence that a HPV E5 protein had transforming activity in mammalian cells, as expression of HPV-6 E5 in established murine fibroblasts lead to anchorage independent growth [23]. Later it was shown that HPV-16 E5 also induces anchorage independence, more efficient growth in low serum and tumorigenic transformation of murine keratinocytes and fibroblasts [62,63,86]. In addition, the acute expression of HPV-16 E5 stimulates cellular DNA synthesis in primary human keratinocytes, and in cooperation with E7, induces proliferation of primary rodent cells [12,102,113,117]. The transforming activity of E5 from HPV-59 and rhesus papillomavirus has been demonstrated in various cell types and assays [38,88].

Mutational analyses of E5 have shown that the protein tolerates many changes without losing transforming activity. This, in conjunction with the small size of E5, indicates that E5 exerts its effects through interaction with, or modulation of, cellular proteins rather than enzymatic activity [49,50,59]. The decreased transforming activity of some of the E5 mutants might thus be due to failure of E5 to interact properly with cellular targets and, as a consequence, to transform cells.

The remainder of this chapter will focus on the cellular targets of E5 and on the role of these interactions in the transformation process.

E5 and growth factor receptors

BPV-1 E5 and PDGF β receptor

Early biochemical studies revealed the BPV-1 E5 cooperates with several PTK growth factor receptors in transforming NIH 3T3 cells [50,66]. The platelet-derived growth factor β-receptor (PDGF-R) is constitutively activated in E5 transformed cells: BPV-1 E5 binds to the PDGF-R and induces receptor dimerization, *trans*-phosphorylation and elevated receptor tyrosine kinase activity [61,71,84,85]. In E5 transformed cells there is a constitutive association between the receptor and phospholipase Cγ, phospho-inositol 3-kinase (PI3-K) and ras GTPase activating protein, SH2 domain-containing cellular proteins that play essential roles in the response to PDGF [24,34,41,61,72,83,84]. Indeed, in E5-transformed cells, a signal transduction complex consisting of the E5 protein, activated PDGF-R and associated signalling molecules can be detected and physically separated from inactive PDGF-R, indicating a true E5-mediated activation [60].

Gene transfer experiments in cell lines lacking endogenous PTK receptors demonstrated that only PDGF β receptor cooperates with E5 and that its activation is required for cell transformation by E5 [34,41,72]. Furthermore, E5 can activate PDGF-R mutants lacking the extracellular ligand binding domain, indicating that E5-induced activation of the receptor is ligand-independent [34]. Inhibition of the PDGF-R tyrosine kinase activity leads to the reversal of the E5 transformed phenotype [58,59]. There is, therefore, a large body of evidence pointing to the critical role of E5-induced PDGF-R activation in cell transformation.

BPV-1 E5 and the PDGF β receptor form stable complexes, in which the two proteins are in opposite orientation and E5 interacts with the transmembrane and juxtamembrane domains of the PDGF-R [25,34,42,81–83,100]. E5 binding induces receptor dimerization which brings the kinase domains of the two monomers into proximity thus starting *trans*-phosphorylation of the receptor [61].

On the basis of extensive mutational analysis, it has been proposed that BPV E5/PDGF-R complex formation requires at least two specific interactions: an electrostatic bond between a juxtamembrane lysine (position 499 in mouse) in the PDGF β receptor and aspartic acid 33 of the BPV-1 E5 protein, and a hydrogen bond between a transmembrane threonine (position 513 in mouse) in the receptor and glutamine 17 of E5 [49,50,57–59]. Mutations of E5 glutamine 17 and aspartic acid 33 abolish cell transformation [70,99] highlighting the importance of these residues and their interaction with the PDGF-R in the transformation process.

In the E5 dimer, the two aspartic acid residues face away from the dimer interface and are thus able to form salt bridges with the lysines on the PDGF β receptor molecules, while the two glutamines are able to form hydrogen bonds not only with the PDGF-R but also across the E5 dimer interface. Each PDGF-R molecule interacts with the glutamine of one E5 monomer and the aspartic acid of the other [57,58,67,104]. Each E5 monomer has an independent PDGF-R binding site which is accessible only after E5 dimerization [2]. For an excellent explanatory diagram of the E5-PDGF-R interaction, see Fig. 7.2 of DiMaio and Mattoon [33].

The outcome of the interaction between BPV-1 E5 and PDGF-R is very similar to that brought about by PDGF. However, PDGF causes receptor dimerization and activation through binding of extracellular domains, whereas E5 binds to the transmembrane and juxtamembrane regions. Moreover, E5 can activate the intracellular precursor of the PDGF-R [34,61]. In agreement, a mutant of the PDGF-R, which fails to traffic to the cell membrane, still cooperates with E5 in inducing cell transformation [100]. It is tempting to speculate that E5 activation of the PDGF-R in different cellular compartments could lead to the recruitment of distinct signal pathways with different effects on the cells.

Despite the importance of PDGF-R binding in E5-induced cell transformation, certain E5 mutants transform cells without binding to or activating the PDGF-R. In particular, the elevated levels of PI3-K in cells transfected with these E5 mutants indicate that under certain conditions E5 can utilise additional signaling pathways such as c-src for activating PI3-K and mediating cell transformation [99,105,105b]. Conversely there are mutant E5 proteins that complex with the PDGF-R and cause receptor tyrosine phosphorylation, but are incapable of cell transformation [67,71].

HPV-16 E5 and EGF receptor

Several lines of evidence suggest that the co-operation between the HPV-16 E5 protein and the receptor tyrosine kinase signalling pathway plays an important role in cell transformation. However, unlike BPV-1 E5, it is the epidermal growth factor receptor (EGF-R) which is involved in HPV E5 transformation. This major difference may reflect the different tissue tropism of the two viruses: HPVs infect exclusively epithelial cells, rich in EGF-R and with no PDGF-R; BPV-1 infects fibroblasts and mesenchymal cells that express high levels of PDGF-R.

HPV-16 E5 or HPV-6 E5a protein induces anchorage independent growth of NIH 3T3 cells [62,86,103]. Cells lacking endogenous EGF-R can be transformed by E5 only when the E5 gene is co-expressed with the EGF-R gene, and colony formation is enhanced by treatment with EGF but not PDGF [86]. E5-transformed cells are more sensitive to growth factor treatment. Several reports indicate that cells transfected with the HPV-16 E5 gene exhibit a more pronounced response to growth factor or phorbol ester treatment, as assessed by the level of transcription of early response genes such as c-fos and c-jun [12,28,62,86]. Also EGF treatment potentiates the mitogenic effect of the HPV-16 E5 protein on keratinocytes in cooperation with the HPV E7 protein. Cells expressing HPV-16 E5 display sustained activation of the signal cascade operating downstream of EGF-R activation: the ras/mitogen-activated protein (MAP) kinase cascade is increased both in response to, and even in the absence of, EGF, and activation of protein kinase C following phorbol ester treatment is enhanced [28,38,46]. Although there are indications that E5 can interact with the EGF pathway without the presence of ligand, the HPV-16 E5 protein primarily affects the function of the EGF-R in the presence of ligand. In monolayer cultures of human keratinocytes transfected by the HPV-16 E5 gene, there are elevated levels of cell surface EGF-R, a result of both reduced ligand-induced degradation of the receptor in endosomes and increased recycling of the EGF-R to the cell surface [102,103]. This is accompanied by increased ligand-induced tyrosine phosphorylation of the receptor [28,103]. Similar effects on EGF-R expression and activation are produced by E5 also in stratified raft cultures of human keratinocytes [108].

From the results described above, it would appear that the primary effect of the HPV E5 protein may be not to activate the receptor directly, but rather to sensitize cells so that they are more responsive to EGF. However interactions with the EGF-R do not appear to account for all the effects of the HPV-16 E5 protein on cellular signal transduction pathways, as it was demonstrated that E5 causes a growth

-independent increase in tyrosine phosphorylation of phospholipase C-γ-1

6 E5 and endothelin-1 receptor

n keratinocytes express endothelin A receptors (ET$_A$R) and produce
ielin-1 (ET-1), which stimulates a growth response in these cells [9]. The effect
autocrine loop on keratinocyte growth is increased in HPV-immortalized
iocytes [116]. In primary keratinocytes transient expression of HPV-16 E5
s increased DNA synthesis and serum-free proliferation in response to
ielin-1. These effects seem not to be linked to an increase in the number or the
ng of endothelin receptor [117]. The ET$_A$R is a G protein-coupled receptor
us very different from tyrosine kinase receptors. E5 therefore is capable of
cting with, and enhancing the signaling of, different classes of growth factor
ors.

spite recent advances in the study of HPV-16 E5, the biochemical basis of E5
ation of various growth factor signaling pathways and consequent cell trans-
tion requires further investigation.

16kDa ductin/subunit c.

5 proteins of both BPVs and HPVs bind a 16kDa cellular protein [26,36,40,44].
ikDa protein is a multifunctional protein with four transmembrane domains
) [37]: as "ductin", it is a component of the connexon, the channel that allows
:llular communication through gap junctions responsible for homeostasis; as
iit c", it is a component of the V0 sector of the vacuolar H$^+$-ATPase
'Pase) which in mammalian cells and in yeast is responsible for the acidifi-
of the endomembrane compartments.

sociation between 16kDa and E5 appears to be mediated largely by trans-
rane interactions [4,42,43]. Mutational analysis has shown that a conserved
ic acid residue within the fourth transmembrane domain of the 16kDa, known
ssential for V-ATPase activity, is critical for binding to glutamine 17 within the
ihobic domain of BPV-1 E5. Noticeably, mutants of the yeast 16kDa in the 143
ic acid residue transform NIH 3T3 cells in a way similar to E5, confirming a
r 16kDa in papillomavirus transformation [4]. As already pointed out, residue
rucial for binding to PDGF-R and for cell transformation; glutamine 17 thus
s to play a major role in the function of BPV-1 E5. Likewise, mutation of the
asparagine residue 17 in BPV-4 E5 affects the transforming ability of the
[76].

p junction intercellular communication (GJIC) contributes to the integration
vidual cells into organised tissues and helps maintain homeostatic control. In
:pressing either BPV or HPV E5, GJIC is inhibited [36,78] and this inhibition

16kDa ductin/subunit c

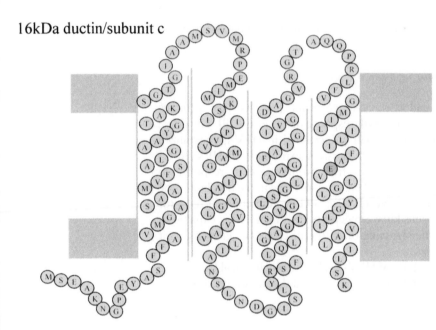

Fig. 2. Secondary structure of bovine 16kDa ductin/subunit c [40]. The glutamate 139 (residue 143 of yeast 16kDa) which mediates binding to E5 [4] is in green. The fourth transmembrane domain binds integrins [97].

43, another component of gap junction [78], and connexin 43 is down-regulated in HPV-16 E5-expressing differentiating keratinocytes, indicating a complex involvement of E5 in cell–cell communication [109]. It is likely that the down-regulation of GJIC by E5 makes transformed cells refractory to growth inhibitory signals originating from neighbouring cells. In support of this hypothesis, closure of cell–cell communication takes place during tumour progression [48].

The V-ATPase complex is responsible for pumping H^+ ions against a gradient in various cellular compartments, including the Golgi apparatus, endosomes, lysosomes, and clathrin-coated vesicles. Acidification of endosomes and Golgi apparatus is impaired in cells expressing HPV or BPV E5, and it has been proposed that E5 directly inhibits V-ATPase activity through its binding to 16kDa subunit c [13,92,102]. Endomembrane alkalinization may be responsible for the inhibition of EGF-R degradation, thereby allowing sustained signaling and increased receptor recycling to the plasma membrane [103,120], while the unbalancing of Golgi pH would affect the trafficking of many important growth regulatory proteins en route to their final destination in the cell (Fig. 3).

In primary transformed cells expressing BPV-4 E5, the Golgi apparatus is swollen and fragmented [7]. This disturbance of Golgi apparatus architecture is due to the elevated pH as a similar disruption is observed in control cells treated with iono-

E5 mechanisms of action

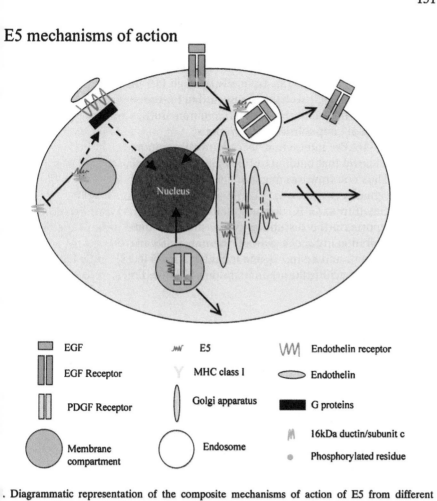

EGF

EGF Receptor

PDGF Receptor

Membrane
compartment

E5

MHC class I

Golgi apparatus

Endosome

Endothelin receptor

Endothelin

G proteins

16kDa ductin/subunit c

Phosphorylated residue

. Diagrammatic representation of the composite mechanisms of action of E5 from different
ɔmaviruses. E5 interacts with, and activates growth factor receptors [33], and prevents transport of
I to the cell membrane [8]. Interaction of E5 with 16kDa ductin/subunit c blocks gap junctions [36],
ʦ acidification of endomembrane compartments [92,102] and impairs endocytic traffic [106]. Open
heads indicate direction of intracellular traffic and full arrowheads indicate signaling pathways.
Broken arrowshafts indicate uncertain pathways.

ʰhanges in Golgi activity, due to the imbalance of pH, would have profound
ɔquences for normal cell functions explaining the pleiotropic action of E5.
ɪbrane receptors and adhesion molecule are glycosylated and transported to
ɪa membrane via the Golgi apparatus [10,90]; small GTPase are post-
lationally modified in the Golgi where they are temporary or permanent
ɪnts [5,24,35,73], and aberrant glycosylation of glycoproteins by Golgi enzymes

contribute to transformation. E5/16kDa subunit c interaction may be responsible also for the reduced motility of mouse fibroblasts expressing E5 [107], since a similar effect is elicited by the V-ATPase inhibitor bafilomycin A. Furthermore E5/16kDa interaction, possibly affecting the movements of integrins from ER through GA to the cell membrane [97], would explain morphological changes observed in E5-transformed cells such as destabilisation of focal adhesions, prominent membrane ruffles and pseudopodes ([8] and M.S. Campo et al., unpublished observations).

Despite the important role of E5–16kDa interaction in cell transformation, mutational analysis of E5 proteins has shown that binding to 16kDa is not sufficient for the induction of transformation. Thus non transforming mutants of E5 can bind 16kDa without enhancing EGF-R signaling or altering V-ATPase function and cellular transformation can be dissociated from GJIC down-regulation [1,7,22,89]. Moreover, there are contradictory reports on the disruption of the multi-subunit V-ATPase complex and cell growth inhibition in *Saccharomyces cerevisiae* following complex formation between 16kDa subunit c and E5 [6,13], and it has been reported that endocytic traffic rather than endosome acidification is impaired in HPV-16 E5 transformed cells [106].

E5 and other cellular targets and functions

E5 and proteins promoting cell proliferation

Members of the c-jun and c-fos families constitute the AP-1 transcription factor, either as jun–jun homodimers or as jun–fos heterodimers. AP-1 is one of the end players in many signal transduction pathways and improper activation and/or over-expression of c-jun/c-fos leads to cell transformation [114]. Expression of c-jun, junB, and c-fos is increased in E5 transformed cells [12,20,21]. This effect is likely to be mediated by a potentiation of the signalling pathway downstream from growth factor receptors as E5-expressing cells treated with EGF have higher levels of c-fos and c-jun RNA. However, increase in c-fos and c-jun expression is also observed in the absence of growth factors, although to a more limited extent, indicating a ligand-independent activation of this pathway by E5 [12,19].

Interestingly, AP-1 binding sites are present in the transcriptional regulatory region (LCR) of papillomavirus, suggesting that E5, by activating AP-1, increases the transcriptional activity of the viral LCR and therefore expression of itself, E6 and E7, in addition to expression of cellular genes. Indeed the HPV-16 LCR is more active in E5 expressing mouse fibroblasts than in control cells [12].

Cyclins and their associated kinases (cdks) regulate the cell cycle and their improper expression/activation induces unscheduled cell proliferation [69,91]. BPV-4 E5 induces increased transcription of the cyclin A gene and elevated expression of cyclin A, accompanied by higher cyclin A-cdk kinase activity [75]. Sustained

E5 and proteins suppressing cell proliferation

The activity of cyclins and cdks is controlled by negative regulators, such as p21^{CIP1} and p27^{KIP1} [91]. HPV E5 proteins repress the transcriptional activity of the p21^{CIP1} gene promoter [110] and promoter repression appears to correlate with cell transformation. As expression of c-jun leads to a reduction of p21^{CIP1} RNA and protein in keratinocytes, the inhibition of p21^{CIP1} expression by E5 is likely due to the observed increase in c-jun.

An effect of BPV-4 E5 transformation is the paradoxically increased levels of p27^{KIP1} [75]. Potentially active p27^{KIP1} is sequestered by an expanded pool of cyclinD1–cdk4 complexes [77] and thus prevented from inhibiting downstream cyclin-cdks and arresting cells in G1. It has been proposed that BPV-4 E5 induces a coordinated increase of p27^{KIP1} and cyclin D1-cdk4, which, together with increased cyclin A-cdk activity, allows continued cell proliferation [77]. Similar findings have been reported for cells expressing c-myc, in which cyclin D-cdk complexes sequester over-expressed p27^{KIP1}, so preventing cell cycle arrest [11,80].

HPV-16 E5 and apoptosis

Many viruses have evolved mechanisms that either block or trigger apoptosis [74]. By blocking apoptosis, viruses prevent premature death of the host cell in order to maximize virus progeny from a lytic infection and facilitate persistent infection. On the other hand virus-induced apoptosis might serve to spread virus progeny while evading host inflammatory responses. Viral proteins are able to block apoptosis by interacting with p53, or through p53-independent pathways.

HPV-16 E5 has been implicated in protecting cells from apoptosis, induced either by UV-B irradiation [121] or by paclitaxel (A. Venuti, unpublished data). UV-B irradiation induces at least four different stress activated pathways including PI3K, ERK1/2MAPK, p38 and JNK signalling. PI3K and ERK1/2MAPK are both downstream from the EGF-R pathway, and EGF-R, PI3K and ERK1/2MAPK are all activated in E5-transformed cells (see above), as is Akt, a downstream effector of PI3K [121]. Inhibition of PI3K or ERK1/2MAPK pathways restores UV-B induced apoptosis, confirming the involvement of these pathways in E5-mediated protection from apoptosis. Contrary to HPV-16 E6, which inhibits apoptosis by p53 inactivation [93,119], E5 prevents apoptosis independently from p53.

Thus E5 may co-operate with E6 in abrogating apoptosis induced by E7 [101] and E2 [31,118] therefore contributing to the promotion of unbalanced cell proliferation and oncogenesis.

Metabolic activity

Phospholipase A activates arachidonic acid metabolism upon growth factor receptor stimulation, thus leading to cyclooxygenase-1 and 2 catalyzed eicosanoid (prostaglandins) biosynthesis. A permanent overactivation of arachidonic acid metabolism appears to be a driving force of tumor development in both experimental animals and man [65]. In monkey and human cell lines expressing BPV-1 E5, phospholipase A activity and arachidonic acid metabolism are increased, and these changes correlate with the transforming ability of E5 [112]. Interestingly, a transformation-defective E5 mutant, also defective for activation of PDGF-R, is still able to activate arachidonic acid metabolism, suggesting an additional role for E5 protein independent of PDGF-R activation.

E5 and MHC class I

Major Histocompatibility Complex class I (MHC I) is responsible for the presentation of antigenic peptides to effector T-cells and therefore plays a critical role in immune surveillance. β2-microglobulin and chaperones, such as TAP, associate with MHC I heavy chain in the endoplasmic reticulum where peptides are loaded onto the MHC I heavy chain in a pH-dependent process [45,87]. The complex is transported from the endoplasmic reticulum through the Golgi apparatus, where dissociation of MHC I from chaperons takes place [115], to the plasma membrane for recognition by T-cells [27].

Given the fact that E5 interferes with the functions of the Golgi apparatus and endosomes [92,102], it was predicted that not only endocytic cellular traffic [106] but also exocytic transport would be disrupted in E5-expressing cells, including transport of the MHC I complex. Indeed, BPV E5 proteins induce down-regulation of MHC I [8]. Down-regulation of MHC I takes place at different levels, including reduced transcription of the MHC I heavy chain gene, lower levels of the MHC I heavy chain protein and impeded transport of the MHC I complex to the cell surface [8]. Lack of surface MHC I is observed also in cells expressing HPV E5 proteins [73b]. It is not yet known how E5 achieves down-regulation of MHC I. Recent results show that in E5-transformed cells, but not in control cells, MHC I heavy chain is sequestered in the Golgi apparatus, and that addition of E5 to control cells also leads to the sequestration of MHC I heavy chain in the Golgi, proof that E5 is the cause of MHC I retention in the Golgi cisternae (M.S. Campo et al., unpublished observations).

Concluding remarks

There is substantial (and growing) evidence that E5 from different PV is involved in cell transformation. Expression of this small hydrophobic protein induces an impressive number of different biological and biochemical effects (Table 1) (Fig. 3), but, given that these results have been obtained not only with different PV E5 proteins, but also in different cell types, it is not possible at the moment to say whether E5 achieves all these effects by interfering with one signal transduction pathway or by intersecting with several. The prevention of acidification of the endomembrane compartment would suggest that malfunction of this compartment is the source of many aspects of cell transformation by E5.

Table 1

The functions of the E5 proteins from BPV-1 and 4, and HPV-16 and 6/11. See text for references. LS, low serum; AI, anchorage independence; CI, contact inhibition.

E5	BPV-1	BPV-4	HPV-16	HPV-6/11
Growth in LS	yes	yes	not known	not known
AI growth	yes	yes	yes	yes
CI abrogation	yes	yes	not known	yes
GJIC down-regulation	yes	yes	yes	not known
GA distortion	yes (in bovine cells)	yes (in bovine cells)	no	not known
EGF-R activation	yes (no ligand required)	not known	yes (ligand required)	not known
PDGF-R activation	yes (no ligand required)	not known	no	not known
Endothelin-R activation	not known	not known	yes	not known
16k interaction	yes	yes	yes	yes
Cyclin A-cdk2 activation	not known	yes	not known	not known
MAP-K activation	not known	not known	yes	yes
JNK activation	not known	not known	yes	yes
PI3K activation	yes	not known	yes	not known
Src activation	yes	not known	not known	not known
GA and endosomes alkalinization	yes	yes	yes	not known
Down-regulation of surface MHC I	yes	yes	yes	yes
Prevention of apoptosis	not known	not known	yes	not known

Although mutational analyses of E5 have pinpointed amino acid residues and domains critical for transformation and for interaction with or activation of cellular proteins [2,57,58,76,81], the many functions of E5 have been genetically dissociated: binding to 16kDa pore protein or PDGF-R from cell transformation and EGF-R activation; activation of MAP kinases from EGF-R activation; activation of PDGF-R from activation of PI3K and c-Src; alteration of arachidonic acid pathway from PDGF-R activation, and down-regulation of GJIC from full cell transformation [1,7,30,89, 98,99,105]. However, other studies have found a very good concordance between cell transformation by BPV-1 E5 and PDGFR activation [58,60].

Notwithstanding the intellectual difficulties posed by E5, a coherent picture of how E5 may transform cells *in vitro* can be drawn from the wealth of experimental results described above: the interaction of E5 with 16kDa ductin/subunit c causes cell transformation by a "pincer movement": inhibition of endomembrane acidification and activation of growth factor receptors lead to inappropriate signalling and unscheduled cell proliferation, while down-regulation of GJIC leads to loss of homeostatic control between neighbouring cells and allows the transformed cell to proliferate (Fig. 3).

The *in vivo* functions of E5 are even less understood than the mechanisms underpinning *in vitro* cell transformation. Nevertheless, on the basis of the expression pattern of E5 and its localisation in the lesions, it is possible to speculate how E5 could contribute to a successful infection by papillomavirus. Following PV infection of basal keratinocytes, E5-induced loss of surface MHC I expression in the infected basal cells (Fig. 3) would prevent presentation of viral antigens to effector T-cells and thus, in addition to other mechanisms of immune avoidance, such as lack of inflammation, would contribute to evasion of immune surveillance. Expression of E5 in the basal layers of the epithelium would lead to sustained cell proliferation to favour virus-infected cells, but extinction of its expression in the more superficial layers would permit cell differentiation and virion production. If E5 expression should proceed beyond early lesional stages, keratinocyte differentiation and immunological removal of infected cells would not take place and the lesion would be at greater risk for neoplastic progression.

Since a lesion results from, and its fate is determined by, the interplay among all viral proteins and between viral and cellular proteins, the early stages of the virus life cycle and of transformation are aspects of PV biology that cannot be elucidated by the sole analysis of E6 and E7, and it is comforting that new investigation on E5 are producing novel results [8,121]. Furthermore E5 may be a target for therapeutic intervention in papillomavirus disease. A recent report shows that immunisation of mice with HPV-16 E5 expressed by a recombinant adenovirus reduced the growth of tumors induced by E5-expressing syngeneic cells [64].

The model of E5 functions described above remains to be validated but provides a framework for future investigations. These will undoubtedly expand our knowledge and understanding not only of the working mechanisms of E5 but also of the basic biology of the virus.

Acknowledgements

A. Venuti gratefully acknowledges the research support provided by the Associazione Italiana per la Ricerca sul Cancro and the Ministero della Sanità. M.S. Campo is supported by Cancer Research UK (formerly Cancer Research Campaign).

References

1. Adam JL, Briggs MW, McCance DJ. A mutagenic analysis of the E5 protein of human papillomavirus type 16 reveals that E5 binding to the vacuolar H+-ATPase is not sufficient for biological activity, using mammalian and yeast expression systems. Virology 2000; 272: 315–325.
2. Adduci AJ, Schlegel R. The transmembrane domain of the E5 oncoprotein contains functionally discrete helical faces. J Biol Chem 1999; 274: 10249–10258.
3. Anderson RA, Scobie L, O'Neil BW, Grindlay GJ, Campo MS. Viral proteins of bovine papillomavirus type 4 during the development of alimentary canal tumours. Vet J, 1997; 154: 69–78.
4. Andresson T, Sparkowski J, Goldstein DJ, Schlegel R. Vacuolar H(+)-ATPase mutants transform cells and define a binding site for the papillomavirus E5 oncoprotein. J Biol Chem 1995; 270: 6830–6837.

5. Apolloni A, Prior IA, Lindsay M, Parton RG, Hancock JF. H-ras but not K-ras traffics to the plasma membrane through the exocytic pathway. Mol Cell Biol 2000; 20: 2475–2487.

6. Ashby ADM, Meagher L, Campo MS and Finbow ME. E5 transforming proteins of papillomaviruses do not disturb the activity of the vacuolar H$^+$-ATPase. J Gen Virol 2001; 82 2353–2362.

7. Ashrafi GH, Pitts JD, Faccini AM, McLean P, O'Brien V, Finbow ME, Campo MS. Binding of bovine papillomavirus type 4 E8 to ductin (16K proteolipid), down-regulation of gap junction intercellular communication and full cell transformation are independent events. J Gen Virol 2000; 81: 689–694.

8. Ashrafi GH, Tsirimonaki E, Marchetti B, O'Brien PM, Sibbet G, Andrew L Campo MS. Down-regulation of surface MHC class I by papillomavirus E5 proteins. Oncogene 2002; 21: 248–259.

9. Bagnato A, Venuti A, Di Castro V, Marcante ML. Identification of the ETA receptor subtype that mediates endothelin induced autocrine proliferation of normal human keratinocytes. Biochem Biophys Res Commun 1995; 209: 80–86.

10. Bellis SL, Newman E, Friedman EA. Steps in integrin beta 1-chain glycosylation mediated by TGFbeta1 signalling through Ras. J Cell Physiol 1999; 181: 33–44.

11. Bouchard C, Thieke K, Maier A, Saffrich R, Hanley-Hyde J, Ansorge W, Reed S, Sicinski P, Bartek J, Eilers M. Direct induction of cyclin D2 by Myc contributes to cell cycle progression and sequestration of p27. EMBO J 1999; 18: 5321–5333.

12. Bouvard V, Matlashewski G, Gu ZM, Storey A, Banks L. The human papillomavirus type 16 E5 gene cooperates with the E7 gene to stimulate proliferation of primary cells and increases viral gene expression. Virology 1994; 203: 73–80.

13. Briggs MW, Adam JL, McCance D. The human papillomavirus type 16 E5 protein alters vacuolar H(+)-ATPase function and stability in *Saccharomyces cerevisiae*. Virology 2001; 280: 169–175.

14. Brown DR, McCloury TL, Woodsand K, Fife KH. Nucleotide sequence and characterization of human papillomavirus type 83, a novel genital Papillomavirus. Virology 1999; 260: 165–172.

15. Bubb V, McCance DJ, Schlegel R. DNA sequence of the HPV-16 E5 ORF and the structural conservation of its encoded protein. Virology 1988; 163: 243–246.

16. Burkhardt A, Willingham M, Gay C, Jeang KT, Schlegel R. The E5 oncoprotein of bovine papillomavirus is oriented asymmetrically in Golgi and plasma membranes. Virology 1989; 170: 334–339.

17. Burnett S, Jareborg N, DiMaio D. Localization of bovine papillomavirus type 1 E5 protein to transformed basal keratinocytes and permissive differentiated cells in fibropapilloma tissue. Proc Natl Acad Sci USA 1992; 89: 5665–5669.

18. Chang JL, Tsao YP, Liu DW, Huang SJ, Lee WH, Chen SL. The expression of HPV-16 E5 protein in squamous neoplastic changes in the uterine cervix J Biomed Sci 2001; 8: 206–213.

19. Chen SL, Tsao LT, Tsao YP. Antisense oligodeoxynucleotides to c-jun inhibits proliferation of transformed NIH 3T3 cells induced by E5a of HPV-11. Cancer Lett 1994; 85: 119–123.

20. Chen SL, Tsao YP, Yang CM, Lin YK, Huang CH, Kuo SW. Differential induction and regulation of c-jun, junB, junD and c-fos by human papillomavirus type 11 E5a oncoprotein J Gen Virol 1995; 76: 2653–2659.

21. Chen SL, Huang CH, Tsai TC, Lu KY, Tsao YP. The regulation mechanism of c-jun and junB by human papillomavirus type 16 E5 oncoprotein. Arch Virol 1996; 141: 791–800.

22. Chen SL, Tsai TC, Han CP, Tsao YP. Mutational analysis of human papillomavirus type 11 E5a oncoprotein. J Virol 1996; 70: 3502–3508.

23. Chen SL, Mounts P. Transforming activity of E5a protein of human papillomavirus type 6 in NIH 3T3 and C127 cells. J Virol 1990; 64: 3226–3233.

24. Choy E, Chiu VK, Silletti J, Feoktistov M, Morimoto T, Michaelson D, Ivanov IE and Philips MR. Endomembrane trafficking of Ras: the CAAX motif targets proteins to the ER and Golgi. Cell 1999; 98, 69–80.

25. Cohen BD, Goldstein DJ, Rutledge L, Vass WC, Lowy DR, Schlegel R, Schiller J. Transformation-specific interaction of the bovine papillomavirus E5 oncoprotein with the platelet-derived growth factor receptor transmembrane domain and the epidermal growth factor receptor cytoplasmic domain. J Virol 1993; 67: 5303–5311.

26. Conrad M, Bubb VJ, Schlegel R. The human papillomavirus type 6 and 16 E5 proteins are membrane-associated proteins which associate with the 16-kilodalton pore-forming protein. J Virol 1993; 67: 6170–6178.

27. Cresswell P, Bangia N, Dick T, Dietrich G. The nature of the MHC class I peptide loading complex. Immunol Rev 1999; 172: 21–28.

28. Crusius K, Auvinen E, Alonso A. Enhancement of EGF- and PMA-mediated MAP kinase activation in cells expressing the human papillomavirus type 16 E5 protein. Oncogene 1997; 15: 1437–1444.

29. Crusius K, Kaszkin M, Kinzel V, Alonso A. The human papillomavirus type 16 E5 protein modulates phospholipase C-gamma-1 activity and phosphatidyl inositol turnover in mouse fibroblasts. Oncogene 1999; 18: 6714–6718.

30. Crusius K, Rodriguez I, Alonso A. The human papillomavirus type 16 E5 protein modulates ERK1/2 and p38 MAP kinase activation by an EGFR-independent process in stressed human keratinocytes. Virus Genes 2000; 20: 65–69.

31. Desaintes C, Goyat S, Garbay S, Yaniv M, Thierry F. Papillomavirus E2 induces p53-independent apoptosis in HeLa cells. Oncogene 1999; 18: 4538–4545.

32. Dennis JW, Granovsky M, Warren CE. Glycoprotein glycosylation and cancer progression. Biochim Biophys Acta 1999; 1473: 21–34.

33. DiMaio D, Mattoon D. Mechanisms of cell transformation by papillomavirus E5 proteins. Oncogene 2001; 20: 7866–7873.

34. Drummond-Barbosa D, Vaillancourt RR, Kazlauskas A, DiMaio D. Ligand-independent activation of the platelet-derived growth factor beta receptor: requirements for bovine papillomavirus E5-induced mitogenic signaling. Mol Cell Biol 1995; 15: 2570–2581.

35. Erickson JW, Zhang C, Kahn RA, Evans T, Cerione RA. Mammalian cdc42 is a Brefeldin A-sensitive component of the Golgi apparatus. J Biol Chem 1996; 271, 26850–26854.

36. Faccini AM, Cairney M, Ashrafi H, Finbow ME, Campo MS, Pitts JD. The bovine papillomavirus type 4 E8 protein binds to ductin and causes loss of gap junctional intercellular communication in primary fibroblasts. J Virol 1996; 70: 9041–9045.

37. Finbow ME, Harrison M, Jones P. Ductin—a proton pump component, a gap junction channel and a neurotransmitter release channel. Bioessays 1995; 17: 247–255.

38. Ghai J, Ostrow RS, Tolar J, McGlennen RC, Lemke TD, Tobolt D, Liu Z, Faras AJ. The E5 gene product of rhesus papillomavirus is an activator of endogenous Ras and

phosphatidylinositol-3'-kinase in NIH 3T3 cells. Proc Natl Acad Sci 1996; USA 93: 12879–12884.

39. Goldstein DJ, Schlegel R. The E5 oncoprotein of bovine papillomavirus binds to a 16 kd cellular protein. EMBO J 1990; 9: 137–145.

40. Goldstein DJ, Finbow ME, Andresson T, McLean P, Smith K, Bubb V, Schlegel R. Bovine papillomavirus E5 oncoprotein binds to the 16K component of vacuolar H(+)-ATPases. Nature 1991; 352, 347–349.

41. Goldstein D, Li W, Wang LM, Heidaran MA, Aaronson SA, Shinn R, Schlegel R, Pierce JH. The bovine papillomavirus type 1 E5 transforming protein specifically binds and activates the beta-type receptor for the platelet-derived growth factor but not other related tyrosine kinase-containing receptors to induce cellular transformation. J Virol 1994; 68: 4432–4441.

42. Goldstein D, Andresson T, Sparkowski JJ, Schlegel R. The BPV-1 E5 protein, the 16 kDa membrane pore-forming protein and the PDGF receptor exist in a complex that is dependent on hydrophobic transmembrane interactions. EMBO J 1992; 11: 4851–4859.

43. Goldstein D, Kulke R, DiMaio D, Schlegel R. A glutamine residue in the membrane-associating domain of the bovine papillomavirus type 1 E5 oncoprotein mediates its binding to a transmembrane component of the vacuolar H(+)-ATPase. J Virol 1992; 66: 405–413.

44. Goldstein D, Toyama R, Dhar R, Schlegel R. The BPV-1 E5 oncoprotein expressed in Schizosaccharomyces pombe exhibits normal biochemical properties and binds to the endogenous 16-kDa component of the vacuolar proton-ATPase. Virology 1992; 190: 889–893.

45. Gromme M, Uytdehaag FG, Janssen H, Calafat J, van Binnendijk RS, Kenter MJ, Tulp A, Verwoerd D, Neefjes J. Recycling MHC class I molecules and endosomal peptide loading. Proc Natl Acad Sci 1999; USA 96: 10326–10331.

46. Gu Z, Matlashewski G. Effect of human papillomavirus type 16 oncogenes on MAP kinase activity. J Virol 1995; 69: 8051–8056.

47. Han R, Cladel NM, Reed CA, Christensen ND. Characterization of transformation function of cottontail rabbit papillomavirus E5 and E8 genes. Virology 1998; 251: 253–263.

48. Holden PR, McGuire B, Stoler A, Balmain A, Pitts JD. Changes in gap junctional intercellular communication in mouse skin carcinogenesis. Carcinogenesis 1997; 18: 15–21.

49. Horwitz BH, Burkhardt AL, Schlegel R, DiMaio D. 44-amino-acid E5 transforming protein of bovine papillomavirus requires a hydrophobic core and specific carboxyl-terminal amino acids. Mol Cell Biol 1988; 8: 4071–4078.

50. Horwitz BH, Weinstat DL, DiMaio D. Transforming activity of a 16-amino-acid segment of the bovine papillomavirus E5 protein linked to random sequences of hydrophobic amino acids. J Virol 1989; 63: 4515–4519.

51. Hsieh CH, Tsao YP, Wang CH, Han CP, Chang JL, Lee JY, Chen SL. Sequence variants and functional analysis of human papillomavirus type 16 E5 gene in clinical specimens. Arch Virol 2000; 145: 2273–2284.

52. Hu X, Pang T, Guo Z, Ponten J, Nister M, Bernard Afink G. Oncogene lineages of human papillomavirus type 16 E6, E7 and E5 in preinvasive and invasive cervical squamous cell carcinoma. J Pathol 2001; 195: 307–311.

160

53. Hwang ES, Nottoli T, DiMaio D. The HPV16 E5 protein: expression, detection, and stable complex formation with transmembrane proteins in COS cells. Virology 1995; 211: 227–233.

54. Jackson ME, Pennie WD, McCaffery RE, Smith KT, Grindlay GJ, Campo MS. The B subgroup bovine papillomaviruses lack an identifiable E6 open reading frame. Mol Carcinogen 1991; 4: 382–387.

55. Jaggar RT, Pennie WD, Smith KT, Jackson ME, Campo MS. Cooperation between bovine papillomavirus type 4 and ras in the morphological transformation of primary bovine fibroblasts. J Gen Virol 1990; 71: 3041–3046.

56. Kell B, Jewers J, Cason J, Best JM. Cellular proteins associated with the E5 oncoprotein of human papillomavirus type 16. Biochem Soc Trans 1994; 22: 333.

57. Klein O, Kegler-Ebo D, Su J, Smith S, DiMaio D. The bovine papillomavirus E5 protein requires a juxtamembrane negative charge for activation of the platelet-derived growth factor beta receptor and transformation of C127 cells. J Virol 1999; 73: 3264–3272.

58. Klein O, Polack GW, Surti T, Kegler-Ebo D, Smith SO, DiMaio D. Role of glutamine 17 of the bovine papillomavirus E5 protein in platelet-derived growth factor beta receptor activation and cell transformation. J Virol 1998; 72: 8921–8932.

59. Kulke R, Horwitz BH, Zibello T, DiMaio D. The central hydrophobic domain of the bovine papillomavirus E5 transforming protein can be functionally replaced by many hydrophobic amino acid sequences containing a glutamine. J Virol 1992; 66: 505–511.

60. Lai CC, Henningson C, DiMaio D. Bovine papillomavirus E5 protein induces the formation of signal transduction complexes containing dimeric activated platelet-derived growth factor beta receptor and associated signaling proteins. J Biol Chem 2000; 275: 9832–9840.

61. Lai CC, Henningson C, DiMaio D. Bovine papillomavirus E5 protein induces oligomerization and trans-phosphorylation of the platelet-derived growth factor beta receptor. Proc Natl Acad Sci USA 1998; 95: 15241–15246.

62. Leechanachai P, Banks L, Moreau F, Matlashewski G. The E5 gene from human papillomavirus type 16 is an oncogene which enhances growth factor-mediated signal transduction to the nucleus. Oncogene 1992; 7: 19–25.

63. Leptak C, Ramon y Cajal S, Kulke R, Horwitz BH, Riese 2nd DJ, Dotto GP, DiMaio D. Tumorigenic transformation of murine keratinocytes by the E5 genes of bovine papillomavirus type 1 and human papillomavirus type 16. J Virol 1991; 65: 7078–7083.

64. Liu DW, Tsao YP, Hsieh CH, Hsieh JT, Kung JT, Chiang CL, Huang SJ, Chen SL. Induction of CD8 T cells by vaccination with recombinant adenovirus expressing human papillomavirus type 16 E5 gene reduces tumor growth. J Virol 2000; 74: 9083–9089.

65. Marks F, Furstenberger G, Muller-Decker K. Metabolic targets of cancer chemoprevention: interruption of tumor development by inhibitors of arachidonic acid metabolism. Recent Results Cancer Res 1999; 151: 45–67.

66. Martin P, Vass W, Schiller JT, Lowy D, Velu TJ. The bovine papillomavirus E5 transforming protein can stimulate the transforming activity of EGF and CSF-1 receptors. Cell 1989; 59: 21–32.

67. Mattoon D, Gupta K, Doyon J, Loll PJ, DiMaio D. Identification of the transmembrane dimer interface of the bovine papillomavirus E5 protein. Oncogene 2001; 20: 3824–3834.

68. Mayer TJ, Meyers C. Temporal and spatial expression of the E5a protein during the differentiation-dependent life cycle of human papillomavirus type 31b. Virology 1998;

248: 208–217.

69. McDonald ER 3rd, El-Deiry WS. Checkpoint genes in cancer. Ann Med 2001; 33: 113–122.

70. Meyer AN, Xu YF, Webster MK, Smith AE, Donoghue DJ. Cellular transformation by a transmembrane peptide: structural requirements for the bovine papillomavirus E5 oncoprotein. Proc Natl Acad Sci USA 1994; 91: 4634–4638.

71. Nilson LA, Gottlieb R, Polack GW, DiMaio D. Mutational analysis of the interaction between the bovine papillomavirus E5 transforming protein and the endogenous beta receptor for platelet-derived growth factor in mouse C127 cells. J Virol 1995; 69: 5869–5874.

72. Nilson LA, DiMaio D. Platelet-derived growth factor receptor can mediate tumorigenic transformation by the bovine papillomavirus E5 protein. Mol Cell Biol 1993; 13: 4137–4145.

73. Noguchi Y, Nakamura S, Yasuda T, Kitigawa M, Kohn LD, Saito Y, Hirai A. Newly synthesised rhoA, not ras, is isoprenylated and translocated to membranes coincident with progression of the G1 to S phase of growth stimulated rat FRTL-5 cells. J Biol Chem 1998; 273: 3649–3653.

73b. O'Brien PM, Campo MS. Immune evasion by papillomavirus. In: Viral Escape from Immune Surveillance (MS Campo, Editor). Virus Res 2002; in press.

74. O'Brien V. Viruses and apoptosis. J Gen Virol 1998; 79: 1833–1845.

75. O'Brien V, Campo MS. BPV-4 E8 transforms NIH3T3 cells, up-regulates cyclin A and cyclin A-associated kinase activity and de-regulates expression of the cdk inhibitor p27Kip1. Oncogene 1998; 17: 293–301.

76. O'Brien V, Ashrafi GH, Grindlay GJ, Anderson R, Campo MS. A mutational analysis of the transforming functions of the E8 protein of bovine papillomavirus type 4. Virology 1999; 255: 385–394.

77. O'Brien V, Grindlay GJ, Campo MS. Cell transformation by the E5/E8 protein of bovine papillomavirus type 4: p27^{Kip1}, elevated through increased protein synthesis is sequestered by cyclin D1-CDK4 complexes. J Biol Chem 2001; 276: 33861–33868.

78. Oelze I, Kartenbeck J, Crusius K, Alonso A. Human papillomavirus type 16 E5 protein affects cell–cell communication in an epithelial cell line. J Virol 1995; 69: 4489–4494.

79. Pennie WD, Grindlay GJ, Carney M, Campo MS. Analysis of the transforming functions of bovine papillomavirus type 4. Virology 1993; 193: 614–620.

80. Perez-Roger I, Kim SH, Griffiths B, Sewing A, Land H. Cyclins D1 and D2 mediate myc-induced proliferation via sequestration of p27(Kip1) and p21(Cip1). EMBO J 1999; 18: 5310–5320.

81. Petti LM, Reddy V, Smith SO, DiMaio D. Identification of amino acids in the transmembrane and juxtamembrane domains of the platelet-derived growth factor receptor required for productive interaction with the bovine papillomavirus E5 protein. J Virol 1997; 71: 7318–7327.

82. Petti L, DiMaio D. Stable association between the bovine papillomavirus E5 transforming protein and activated platelet-derived growth factor receptor in transformed mouse cells. Proc Natl Acad Sci USA 1992; 89: 6736–6740.

83. Petti L, DiMaio D. Specific interaction between the bovine papillomavirus E5 transforming protein and the beta receptor for platelet-derived growth factor in stably transformed and acutely transfected cells. J Virol 1994; 68: 3582–3592.

162

84. Petti L, Nilson L, DiMaio D. Activation of the platelet-derived growth factor receptor by the bovine papillomavirus E5 transforming protein. EMBO J 1991; 10: 845–855.

85. Petti LM, Ray FA. Transformation of mortal human fibroblasts and activation of a growth inhibitory pathway by the bovine papillomavirus E5 oncoprotein. Cell Growth Differ 2000; 11: 395–408.

86. Pim D, Collins M, Banks L. Human papillomavirus type 16 E5 gene stimulates the transforming activity of the epidermal growth factor receptor. Oncogene 1992; 7: 27–32.

87. Reich Z, Altman JD, Boniface JJ, Lyons DS, Kozono H, Ogg G, Morgan C, Davis MM. Stability of empty and peptide-loaded class II major histocompatibility complex molecules at neutral and endosomal pH: comparison to class I proteins. Proc Natl Acad Sci USA 1997; 18: 2495–2500.

88. Rho J, de Villiers EM, Choe J. Transforming activities of human papillomavirus type 59 E5, E6 and E7 open reading frames in mouse C127 cells. Virus Res 1996; 44: 57–65.

89. Rodriguez MI, Finbow ME, Alonso A. Binding of human papillomavirus 16 E5 to the 16 kDa subunit c (proteolipid) of the vacuolar H+-ATPase can be dissociated from the E5-mediated epidermal growth factor receptor overactivation. Oncogene 2000; 19: 3727–3732.

90. Rudd PM, Wormaid MR, Stanfield RL, Huang M, Mattson N, Speir JA, DiGennaro JA, Fetrow JS, Dwek RA, Wilson IA. Roles for glycosylation of cell surface receptors involved in cellular immune recognition. J Mol Biol 1999; 293: 351–366.

91. Sandhu C, Slingerland J. Deregulation of the cell cycle in cancer. Cancer Detect Prev. 2000; 24: 107–118.

92. Schapiro F, Sparkowski J, Adduci A, Suprynowicz F, Schlegel R, Grinstein S. Golgi alkalinization by the papillomavirus E5 oncoprotein. Cell Biol 2000; 148: 305–315.

93. Scheffner M, Werness BA, Huibregtse JM, Levine AJ, Howley P. The E6 oncoprotein encoded by human papillomavirus types 16 and 18 promotes the degradation of p53. Cell 1990; 63: 1129–1136.

94. Schiller JT, Vass WC, Vousden KH, Lowy DR. E5 open reading frame of bovine papillomavirus type 1 encodes a transforming gene. J Virol 1986. 57: 1–6.

95. Schlegel R, Wade-Glass M, Rabson MS, Yang YC. The E5 transforming gene of bovine papillomavirus encodes a small, hydrophobic polypeptide. Science 1986. 233: 464–467.

96. Settleman J, Fazeli A, Malicki J, Horwitz BH, DiMaio D. Genetic evidence that acute morphologic transformation, induction of cellular DNA synthesis, and focus formation are mediated by a single activity of the bovine papillomavirus E5 protein. Mol Cell Biol 1989; 9: 5563–5572.

97. Skinner MA, Wildeman AG. β(1) integrin binds the 16-kDa subunit of vacuolar H(+)-ATPase at a site important for human papillomavirus E5 and platelet-derived growth factor signaling. J Biol Chem 1999; 274: 23119–23127.

98. Sparkowski J, Anders J, Schlegel R. E5 oncoprotein retained in the endoplasmic reticulum/cis Golgi still induces PDGF receptor autophosphorylation but does not transform cells. EMBO J 1995; 14: 3055–3066.

99. Sparkowski J, Mense M, Anders J, Schlegel R. E5 oncoprotein transmembrane mutants dissociate fibroblast transforming activity from 16-kilodalton protein binding and platelet-derived growth factor receptor binding and phosphorylation. J Virol 1996; 70: 2420–2430.

100. Staebler A, Pierce JH, Brazinski S, Heidaran MA, Li W, Schlegel R, Goldstein DJ. Mutational analysis of the beta-type platelet-derived growth factor receptor defines the site of interaction with the bovine papillomavirus type 1 E5 transforming protein. J Virol 1995; 69: 6507–6517.

101. Stoppler H, Stoppler MC, Jhonson E, Simbulan-Rosenthal CM, Smulson ME, Iyer S, Rosenthal DS, Schlegel R. The E7 protein of human papillomavirus type 16 sensitizes primary human keratinocytes to apoptosis. Oncogene 1998; 17: 1207–1214.

102. Straight SW, Herman B, McCance DJ. The E5 oncoprotein of human papillomavirus type 16 inhibits the acidification of endosomes in human keratinocytes. J Virol 1995; 269: 3185–3192.

103. Straight SW, Hinkle PM, Jewers RJ, McCance DJ. The E5 oncoprotein of human papillomavirus type 16 transforms fibroblasts and effects the downregulation of the epidermal growth factor receptor in keratinocytes. J Virol 1993; 67: 4521–4532.

104. Surti T, Klein O, Ascheim K, DiMaio D, Smith SO. Structural models of the bovine papillomavirus E5 protein. Proteins 1998; 33: 601–612.

105. Suprynowicz FA, Sparkowski J, Baege A, Schlegel R. E5 oncoprotein mutants activate phosphoinositide 3-kinase independently of platelet-derived growth factor receptor activation. J Biol Chem 2000; 275: 5111–5119.

105b.Suprynowicz FA, Baege A, Sunitha I, Schlegel R. c-Src activation by the E5 oncoprotein enables transformation independently of PDGF receptor activation. Oncogene 2002; 21: 1695–1706.

106. Thomsen P, van Deurs B, Norrild B, Kayser L. The HPV16 E5 oncogene inhibits endocytic trafficking. Oncogene 2000; 19: 6023–6032.

107. Thomsen P, Rudenko O, Berezin V, Norrild B. The HPV-16 E5 oncogene and bafilomycin A(1) influence cell motility. Biochim Biophys Acta 1999; 1452: 285–295.

108. Tomakidi P, Cheng H, Kohl A, Komposch G, Alonso A. Modulation of the epidermal growth factor receptor by the human papillomavirus type 16 E5 protein in raft cultures of human keratinocytes. Eur J Cell Biol 2000; 79: 407–412.

109. Tomakidi P, Cheng H, Kohl A, Komposch G, Alonso A. Connexin 43 expression is downregulated in raft cultures of human keratinocytes expressing the human papillomavirus type 16 E5 protein. Cell Tissue Res 2000; 301: 323–327.

110. Tsao YP, Li LY, Tsai TC, Chen SL. Human papillomavirus type 11 and 16 E5 represses p21(WafI/SdiI/CipI) gene expression in fibroblasts and keratinocytes. J Virol 1996; 70: 7535–7539.

111. Ullman CG, Haris PI, Kell B, Cason J, Jewers RJ, Best JM, Emery VC, Perkins SJ. Hypothetical structure of the membrane-associated E5 oncoprotein of human papillomavirus type 16. Biochem Soc Trans 1994; 22: 439S.

112. Vali U, Kilk A, Ustav M. Bovine papillomavirus oncoprotein E5 affects the arachidonic acid metabolism in cells. Int J Biochem Cell Biol 2001; 33: 227–235.

113. Valle FG, Banks L. The human papillomavirus (HPV)-6 and HPV-16 E5 proteins co-operate with HPV-16 E7 in the transformation of primary rodent cells. J Gen Virol 1995; 76: 1239–1245.

114. van Dam H, Castellazzi M. Distinct roles of Jun: Fos and Jun: ATF dimers in onco-genesis. Oncogene 2001; 20: 2453–2464.

115. van Leeuwen JEM, Kearse KP. Deglucosylation of N-linked glycans is an important step in the dissociation of calreticulin-class I-TAP complexes. Proc Natl Acad Sci USA 1996;

164

93: 13997–14001.

116. Venuti A, Marcante ML, Flamini S, Di Castro V, Bagnato A. The autonomous growth of human papillomavirus type 16-immortalized keratinocytes is related to the endothelin-1 autocrine loop. J Virol 1997; 71: 6898–6904.

117. Venuti A, Salani D, Poggiali F, Manni V, Bagnato A. The E5 oncoprotein of human papillomavirus type 16 enhances endothelin-1-induced keratinocyte growth. Virology 1998; 248: 1–5.

118. Webster K, Parish J, Pandya M, Stern PL, Clarke AR, Gaston K. The human papillomavirus (HPV) 16 E2 protein induces apoptosis in the absence of other HPV proteins and via a p53-dependent pathway J Biol Chem 2000; 275: 87–94.

119. Werness BA, Levine AJ, Howley PM. Association of human papillomavirus types 16 and 18 E6 proteins with p53. Science 1990; 248: 76–79.

120. Yoshimori T, Yamamoto A, Moriyama Y, Futai M, Tashiro Y. Bafilomycin A1, a specific inhibitor of vacuolar-type H(+)-ATPase, inhibits acidification and protein degradation in lysosomes of cultured cells. J Biol Chem 1991; 266: 17707–17712.

121. Zhang B, Spandau DF, Roman A. E5 protein of human papillomavirus type 16 protects human foreskin keratinocytes from UVB-irradiation-induced apoptosis. J Virol 2002; 76: 220–231.

Human Papillomaviruses
D.J. McCance (editor)
© 2002 Elsevier Science B.V. All rights reserved

Human papillomavirus immunology and vaccine development

Robert Rose

Departments of Medicine, Microbiology and Immunology, and the David H. Smith Center for Vaccine Biology and Immunology, University of Rochester School of Medicine and Dentistry, 601 Elmwood Avenue, Rochester, New York 14642, USA

Introduction

Uterine cervical carcinoma is the second most common malignancy among women worldwide, and is the most common female cancer in developing regions [53]. Infection with certain mucosatropic human papillomaviruses (HPVs) is now known to be a necessary prerequisite for development of this disease [22]. Anogenital HPV infection in susceptible populations is characterized by high rates of acquisition and spontaneous clearance [74], suggesting that most infected individuals can mount an effective response to infection. However, the mechanisms underlying such responses remain to be characterized. Encouraging progress has been made toward the development of preventive vaccines, and efforts are underway to design improved methods for treatment of established disease. This chapter examines recent developments in HPV immunologic research, and discusses issues relevant to the design and implementation of effective strategies for mass immunization against HPV-associated disease.

Background

The development of sensitive molecular methods in the mid-1970s led to the recognition that "human wart virus" actually comprised a diverse group of related viruses with variable tissue tropisms and disease associations [52,107]. Serologic studies conducted prior to that time are difficult to interpret, as the antigens used often consisted of virions pooled after recovery from lesions that occurred at multiple anatomic sites. Nevertheless, information from such studies may be useful. For example, it was reported that individuals with cutaneous warts often produce virion-specific antibodies, and that spontaneous regression is observed less frequently in patients that lack such responses, or in patients with IgM antibodies alone [126,143]. Also reported was the observation that patients with chronic or multiple warts are less likely to have IgG against whole virions [92], as are patients whose warts are devoid of virions [126].

Following recognition of papillomavirus genetic diversity, investigators began to use virions recovered from defined tissue sources for immunologic studies, and thus generated potentially more useful information. Using native virions of HPV type 1 (HPV-1), for example, it was shown that virion-specific IgG antibodies could be detected at high frequency in non-selected populations [109,130], and that such antibodies were immunoreactive primarily with non-linear (i.e., conformational) epitopes of the viral capsid [130]. Variations in capsid antigenicity were observed in studies that utilized virions recovered from different anatomic sites. Viac et al. [139] reported that antibodies reactive with plantar wart virions are detected more commonly in sera from patients with plantar warts than in sera from patients with either common, flat, anogenital, or laryngeal warts. Similarly, using immunoelectron microscopy Anisimová et al. showed that HPV-1 and HPV-2 virions are antigenically distinct [8]. These observations suggested that virions recovered from different anatomic sites possessed unique antigenic properties, and thus might be useful antigens for immunologic studies.

For many years the study of immune responses to anogenital HPV infection was made difficult because of a general inability to obtain native virions, either by propagation *in vitro* or by recovery from infected tissues. With the development in 1985 of a human xenograft immunodeficient mouse system for propagation of anogenital HPV type 11 (HPV-11) [76,77], researchers gained a useful means of producing native virions for serology [14,15,17,37] and for vaccine-related studies [18,35,36]. A limitation of this system, i.e., the ability to propagate virions of only a relative few anogenital types, was later circumvented to some extent by the production of non-infectious empty capsids, or virus-like particles (VLPs), through recombinant expression of the papillomavirus L1 major capsid protein [58,68,117, 142,152]. VLPs closely approximate the immunologic properties of native virions [34,70,114–116,119,137], and thus represent an excellent alternative source of antigen for serologic investigations of anogenital HPV disease.

Humoral responses in anogenital HPV infection

Since their description roughly a decade ago, VLPs have become indispensable tools for investigating immune responses in natural infection, for HPV epidemiologic studies, and for vaccine development [67,122]. VLPs form spontaneously following recombinant expression of the papillomavirus L1 major capsid protein in insect (Sf9 cells), or yeast cells [58,68,117], and can incorporate the L2 minor capsid protein when these proteins are co-expressed in the same system [71]. VLPs closely approximate the antigenicity of native anogenital virions, and thus are good antigens for serologic purposes. In a comparison of HPV-11 virions and VLPs, for example, excellent correlation of sample seroreactivities ($r = 0.87$) was obtained when sera from patients with anogenital warts were tested by ELISA against these antigens [15,119].

The first study of capsid seroresponses in anogenital HPV disease utilized HPV-11 virions propagated in the xenograft system. In that study capsid seropositivity was found to correlate with disease; however, responses were not detected in all infected individuals [14]. Results from subsequent studies of virion/VLP seropositivity have been consistent with this observation [15,17,29,70, 119]. For example, in an HPV-16 VLP enzyme-linked immunosorbent assay (ELISA) [71] HPV-16 capsid antibodies were detected in sera from a majority of women infected with HPV-16 [70]. Using a vaccinia virus system, Carter et al. reported detection of HPV-1 capsid antibodies in 36% of sera from unselected individuals [29]. In a more recent study of seroresponses to incident infection, Carter et al. reported slight variations in seroconversion rates against three anogenital HPV types; however, all rates fell eventually within a relatively narrow range (i.e, 54 to 69% seropositivity within 18 months of infection) [30]. Also in that study certain differences were noted in seroconversion patterns among the types examined (i.e., types 6, 16 and 18). Specifically, HPV-6 seroconversion usually coincided with detection of HPV-6 DNA, whereas seroconversion to HPV-16 or HPV-18 was more often delayed by approximately 6 to 12 months following DNA detection. Genotype-specific differences were also noted with respect to persistence of sero-positivity, with responses in most (i.e., two-thirds) individuals with incident HPV-6 infection becoming undetectable at follow-up, while in most individuals (i.e., 75%) with HPV-16 or 18 incident infection, seropositive status tended to persist through follow-up [30].

Presence of capsid-specific serum antibodies prior to exposure has been correlated with immunity from infectious challenge [133]; however, a role for capsid antibodies in natural infection has not been defined. An association between capsid seropositivity and persistence of viral DNA has been reported [30,31,44,149], but there is little evidence of a correlation between the presence or absence of such responses and stage of disease [1,42]. Oncogenic HPV capsid seropositivity can be detected, for example, in invasive disease [145]. There is evidence to suggest that systemic, but not local, capsid-specific immunoglobulin A (IgA) responses may correlate with disease resolution [20], suggesting the possibility that VLP-based assays may be useful for monitoring disease activity.

Seroresponses to other viral proteins have been examined in the context of specific diseases. In a study of individuals with varying stages of cervical intra-epithelial neoplasia (i.e., CIN grades I, II or III, or invasive disease), sera were evaluated in an ELISA for E2 serum IgA antibody responses [113]. Results indicated that elevated E2 serum IgA could be detected in mild or intermediate disease (i.e., CIN I-II) relative to normal controls, with lower levels in cervical cancer patients, suggesting the possible existence of an inverse relationship between E2 serum IgA levels and disease progression [113]. Serum antibodies against E6 and E7 proteins of HPV types 16 and 18 have been associated with cervical cancer [13,141]. Elevated levels of E6 and E7 antibodies have also been found in cervical lavage specimens recovered from women with invasive disease versus healthy controls [136]. The significance of these observations is unclear at present; however, given the context, such responses may be reflective of an inappropriate and/or ineffective immune response to infection.

Mucosal antibodies are thought to play an important role in immune defense against sexually transmitted pathogens, including HPV [86]. Mucosal immunoglobulin A (IgA) antibody responses have been detected at low frequency in HPV natural infection [57,82,112] and with greater consistency in experimental animals following mucosal administration of VLPs [11,47,50,55,103]. Wang et al. found, for example, that cervical IgA antibody responses are frequently made by women with cervical intraepithelial neoplasia, and are particularly evident in association with low-grade lesions (i.e., CIN I) [146]. In a recent study of college-aged women examined at four-month intervals over a 3–4 year period, cervical IgA responses were detected in a minority (11%) of women, and were not associated with HPV-16 DNA detection [57]. However, in that study detection of secretory IgA was associated with prior detection (i.e., within the previous 4–8 months) of squamous intraepithelial lesions [57]. In an evaluation of HPV-16 capsid antibody responses in saliva, Marais et al. reported detection of VLP-specific IgA in most (44/81; 54.3%) women with cervical HPV disease (i.e., 75 women with CIN, and 6 women with squamous cell carcinoma of the cervix) [82]. Interestingly, VLP-specific IgA antibodies were detected more frequently in saliva than in serum (54% vs 34%), whereas VLP-specific IgG was detected more frequently in serum rather than in saliva (68% vs 34%) [82]. Although the significance of this observation remains to be established, the results suggest the potential utility of evaluating oral secretions by ELISA as a convenient non-invasive screening method.

Cellular immune responses

Observations made in clinical settings have indicated that cellular immune mechanisms are key elements of effective responses to anogenital HPV infection [123]. For example, an increased incidence of anogenital HPV disease has been observed in patients with primary or secondary cellular immunodeficiencies [59,128,129]. Despite significant effort, no consistent patterns have emerged concerning T helper (Th) and/or cytotoxic T lymphocyte (CTL) specificities that may be associated with efficient clearance of infection. Earlier studies, conducted with whole virions, recombinant vaccinia viruses, or synthetic peptides, examined potentially operative mechanisms. For instance, delayed-type hypersensitivity (DTH) responses to HPV antigen preparations have been detected in patients with warts, compared with control subjects [66,140]. Synthetic peptides were used successfully to identify T-cell lymphoproliferative epitopes in HPV-16 E6 and L1 [132], and in HPV-1 E4 [131]. Cellular responses to HPV-16 and BPV proteins have been studied in mice in vaccinia virus systems, and immunizations with vaccinia virus recombinants expressing HPV-16 L1 [153], E6 [87], or E7 [32,87] proteins were found to promote tumor regression, presumably through induction of cytotoxic T lymphocyte (CTL) responses. Similar responses were observed in the bovine system using vaccinia virus vectors [88] or bacterially-expressed fusion proteins [26,63]. Interestingly, Campo et al. [26] reported that cattle immunized with BPV L2 or E7 bacterial fusion proteins either before or after experimental challenge, were equally able to resolve infection more quickly than non-vaccinated animals [26]. While such results suggest the potential of immunization therapy to hasten disease resolution, the mechanisms involved are not yet known.

In patients with anogenital warts, resolution appears to involve an active cellular response. Regressing lesions have been found, for example, to contain greater numbers of T lymphocytes and macrophages than non-regressing lesions [38]. In addition, CD4+ lymphocytes seem to predominate in regressing lesions, and to exhibit signs of lymphocyte activation [38], implying that immune mechanisms characteristic of a delayed type hypersensitivity reaction to foreign antigen may be important for viral clearance [38].

In a study of CTL reactivity against selected peptides from the HPV-16 E4, E6, E7 and L1 open reading frames, higher rates of reactivity to E6 were found in sexually active women without disease versus women with CIN [99]. Similarly, CTL responses against HPV-16 E6 or E7 were more common in HPV-16-positive women without disease than in HPV-16-positive women with CIN [100] and such responses typically involved CD4+ and CD8+ T lymphocytes [101]. More recent evidence suggests that lack of E6-specific CTL responses may contribute to the establishment of persistent disease [102]. Consistent with this possibility, E7-specific CD4+ activity appears to decrease with resolution, and to increase in the context of persistent infection and CIN III [43]. Potentially operative T helper and CTL epitopes have also been identified in the viral E2 protein [19,40,75,81], however, the role of E2-, E6- and/or E7-specific Th and/or CTL in natural disease and immunity remain to be clarified.

Cytokine responses

Recognition that specific cytokine response patterns are elicited following antigen stimulation [93] has generated greater understanding of ways in which pathogens are able to interfere with mechanisms of immune-mediated clearance [90,110]. Not unexpectedly, Th1 cytokine response patterns have been detected in resolving anogenital HPV lesions [124], whereas patterns more consistent with a Th2 response have been observed in persistent disease [6,108]. The suggestion that it may be possible to switch an inappropriate response to one that is potentially more effective [98] provides encouragement for further investigation of this interesting and potentially critical component of natural immunity, and supports the evaluation of cytokine/adjuvant-based immunotherapies potentially capable of altering established patterns of response.

A role for Th1 vs Th2 responses in cervical HPV infection has not yet been established; however, there is evidence that Th1 responses may be important for viral clearance. For example, in a longitudinal study, cytokine expression in cervical mucosa was examined in HPV-positive patients that ultimately cleared infection [124]. In that study each of the individuals examined ($N = 7$) demonstrated a Th1 cytokine expression pattern (i.e., presence of IFN-γ and absence of IL-4) prior to clearance. By contrast, cytokine expression patterns were variable in normal, healthy HPV-negative, control women [124].

The possible association of Th2 response patterns with persistence of disease has also been examined [6]. Cytokine expression patterns (i.e., IL-2 and IFN-γ (Th1), versus IL-4 and IL-6 (Th2)) were evaluated by immunohistochemistry of cervical biopsy specimens collected from healthy individuals and from individuals with severe cervical disease. Results indicated significantly lower densities of IL-2+ cells in biopsies from women with severe disease versus normal individuals, whereas densities of IL4+ cells were found to be elevated in tissues from women with cervical disease compared with histologically normal tissues. Elevated densities of Th2-like cells in biopsies from women with disease were associated with HLA-DR (i.e., Major Histocompatibility Complex Class II) expression by keratinocytes, and with lower densities of intraepithelial Langerhans' cells, suggesting the establishment of a Th2 profile of cytokine responses in persistent infection [6].

HPV Immunoprophylaxis

Parenteral vaccines

Current strategies for HPV immunoprophylaxis rely on the concept of antibody-mediated neutralization of viral infectivity. This possibility was suggested long ago by Segre et al., who demonstrated the ability of convalescent serum to protect against infectious challenge [125]. More recently, virions recovered from HPV-11-infected human xenografts [76,77] were used to generate polyclonal antisera that were found to efficiently neutralize HPV-11 viral infectivity [18,35]. This neutralizing effect was correlated with highly immunogenic non-linear epitopes of the virion [36], and was reproduced by using polyclonal antisera generated against HPV-11 VLPs [34,119]. Other investigators produced VLPs of Bovine Papillomavirus type 1 (BPV-1), which were used to generate polyclonal antibodies that efficiently blocked BPV-1 virion-mediated focus formation in a mouse fibroblast assay [48,68]. With production of VLPs of multiple anogenital HPV genotypes [116,119], it was found that HPV-11 capsid neutralizing domains were antigenically distinct from qualitatively similar capsid epitopes of other prevalent anogenital HPV genotypes (i.e., types 16 and 18), suggesting that such epitopes varied essentially with genotype [116] (Fig. 1). This observation was confirmed by Roden et al., who extended the number of genotypes examined, and also demonstrated low-level capsid antigenic cross-reactivity between certain closely related genotypes [115]. Consistent with these observations, others demonstrated efficient neutralization of HPV-16 virions with antibodies raised against VLPs of HPV-16 and, to a lesser extent, with antibodies raised against VLPs of closely related HPV-31 [16,148]. Taken together, these observations support the concept of antibody-mediated virus neutralization, and demonstrate the ability of VLPs to induce such responses. They also support the use of multivalent VLP vaccine formulations for the induction of protective immunity against multiple anogenital HPV genotypes [51,121].

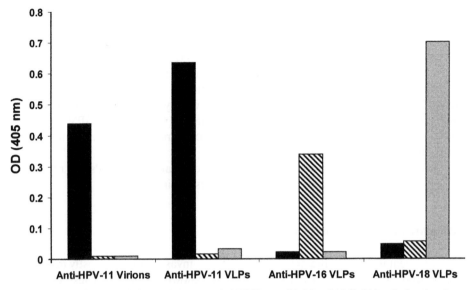

Fig. 1. Capsid antigenic variation among anogenital HPV types 11, 16 and 18. Rabbit polyclonal antisera were raised against HPV-11 virions, HPV-11 L1 VLPs, HPV-16 VLPs or HPV-18 VLPs, and evaluated in an enzyme-linked immunosorbent assay (ELISA) against VLPs of HPV-11 (black bar), HPV-16 (striped bar) or HPV-18 (gray bar). HPV-11 virion and HPV-11 VLP polyclonal antisera were characterized previously and found to efficiently neutralize infectious HPV-11 virions [18,119] (Figure adapted from Ref. [116]).

Studies performed in experimental animals have also provided rationale for the evaluation of VLPs as a vaccine in human subjects. For example, VLP immunizations in rabbits, dogs and cattle have been shown to confer protection from challenge with homologous virus [24,69,133]. Such protection was correlated with non-linear geno-type-restricted capsid epitopes, and immunity from challenge could be conferred to naïve animals via transfer of post-immune serum obtained from VLP-immunized animals, thus supporting the view that protection may be mediated by antibodies [133].

Blinded, randomized, placebo-controlled, dose escalation studies of VLPs have been performed in humans, and results have indicated that VLPs are safe, well tolerated, and induce high-titer serum neutralizing antibody responses [49,60]. In a phase I study of HPV-16 VLPs performed in healthy adults, VLP vaccine was administered by parenteral (intramuscular) injection in two dose levels (i.e., 10 μg or 50 μg), with or without adjuvant (i.e., alum or MF59) [60]. Investigators reported that VLPs were safe and well-tolerated, and induced serum neutralizing antibody responses in all recipients that received the higher antigen dose level, without adjuvant. Modest increases in titer were also observed in high-dose recipients of VLPs co-administered with MF59. VLP ELISA titers and neutralization titers closely correlated, thus supporting the use of VLP ELISA titer as a surrogate for neutralization titer [60].

Similarly, HPV-11 VLPs formulated in aluminum hydroxide were administered by intramuscular injection in healthy adults [49]. In that study, VLPs were administered in four dose levels (i.e., 3 μg, 9 μg, 30 μg and 100 μg), and high serum neutralizing titers were obtained at each dose level tested. Results obtained with an intermediate dose level (i.e., 30 μg) were consistent with results reported by Harro et al., which were obtained with a comparable antigen dose level (i.e., 50 μg) [60]. HPV-11 VLP immunization was also associated with lymphoproliferative responses against homologous and heterologous VLP antigens (i.e., VLPs of HPV types 11, 6 and 16), suggesting the possibility that T cell epitopes may be conserved among these anogenital HPV genotypes [49].

These encouraging results support further evaluation of VLPs for efficacy in preventing HPV-associated disease. Such studies are in progress.

Alternative vaccination strategies

There is great interest in the development of non-invasive and less costly methods of vaccination. Most (~80%) cervical cancer is known to occur in less developed regions [53,96], where use of parenterally administered vaccines generally is impractical. Moreover, protection from oncogenic HPVs and other sexually transmitted pathogens is likely to depend to some extent on immunity acting at genital mucosal surfaces [97]. Mucosal vaccines generally are less expensive to produce, and often can be delivered by unskilled personnel without costly equipment. Importantly, mucosal immunization generally is superior to parenteral vaccination for the induction of mucosal immune responses [86], and mucosal vaccination at a convenient site (e.g., nasal or gastrointestinal mucosa) can elicit a response at one or more distal sites, such as genital or rectal mucosa [89,97]. Adjuvants are usually required, however, to boost mucosal immunogenicity and to prevent the induction of tolerance [86].

Cholera toxin (CT), *Escherichia coli* heat-labile enterotoxin (LT), and their mutant derivatives are promising mucosal adjuvants for co-administered protein antigens [28]. For obvious reasons, CT or LT holotoxins are unlikely to be approved for use in humans. However, LT mutants have been described in which adjuvanticity has been dissociated from toxicity. For example, LT mutant R192G (LT(R192G)) was constructed by site-directed mutagenesis to introduce a single amino acid subst-itution in the active (A) subunit [45]. This mutation rendered the toxin insensitive to trypsin activation and thus greatly diminished toxicity without altering adjuvanticity of the native molecule. LT(R192G) has been evaluated in several studies and found to be an effective mucosal adjuvant [27,33,56].

Synthetic oligodeoxynucleotides containing unmethylated CpG dinucleotide motifs (CpG DNA) represent another class of mucosal adjuvant [78,91] that has demonstrated potent immunostimulatory properties [78,85,144]. CpG oligos have been shown to activate innate immune mechanisms by interacting with mammalian toll-like receptor 9 (TLR9), which may have evolved for the purpose of responding to the presence of bacterial DNA (which contains CpG dinucleotides in unmethylated form) [61].

Oral or intranasal vaccination routes are attractive, particularly for use in resource-poor settings, as such vaccines generally are non-invasive and safe, and can be distributed by relatively unskilled personnel without the need for costly injection equipment. As an example, efforts to eradicate poliovirus have been facilitated greatly by the availability of an oral poliovirus vaccine [46].

Intranasal immunization

Serum and mucosal immune responses can be elicited by VLP intranasal administration [11,47]. Vaginal IgA antibody responses were detected, for example, following intranasal immunization of mice with HPV-16 VLPs co-administered with cholera toxin (CT) [11,47]. Three intranasal immunizations with 5 μg of VLPs administered at weekly intervals induced high durable titers of HPV-16 VLP IgA and IgG in saliva and genital secretions. Co-administration of VLPs with CT was found to enhance the VLP-specific antibody response ten-fold in serum and to a lesser extent in saliva and genital secretions. However, VLPs were found to be only poorly immunogenic when administered by an oral route with or without CT [11]. Dupuy et al. [47] also reported induction of HPV-16 VLP serum IgG and vaginal IgA antibodies following intranasal administration of VLPs with CT, and in that study VLPs administered intranasally without CT were found to be only poorly immunogenic [47]. HPV-6 VLPs have also been evaluated with regard to intranasal immunogenicity [55]. HPV-6 VLPs are also immunogenic, and co-administration of a genetically modified *E. coli* heat-labile enterotoxin (i.e., LTR72) was associated with greater induction of mucosal VLP IgA antibody responses [55]. Phase I studies are in progress to evaluate the safety, tolerability and immunogenicity of VLPs as an intranasal immunogen (D. Nardelli-Haefliger, personal communication).

Oral immunization

VLPs of anogenital HPV types are immunogenic in mice when administered orally with or without adjuvant, and induce systemic and mucosal virus-neutralizing antibody responses [50,118]. Intestinal antigens are thought to enter gut-associated lymphoid tissue (GALT) via M cells located in the Peyer's Patch (PP) epithelium [105]. Several pathogens have been shown to utilize M cells as a portal of entry [7,25,62,65,127,150]; it is thought that induction of systemic responses after oral delivery may rely on cellular binding activity [41], and particularly may be related to the ability of certain types of antigen to bind glycolipids or glycoproteins on intestinal mucosa [41]. Papillomavirus VLPs are able to bind to a variety of cell types [94], and this ability may be relevant to VLP activation of immune responses in gut mucosal tissues [50]. As with live viral or bacterial pathogens, VLPs may be delivered directly to professional antigen presenting cells by M cells underlying the PP epithelium. VLPs of other viral pathogens (e.g., Rotavirus, Norwalk Virus) have also demonstrated an ability to elicit mucosal responses in mice following oral administration [10,106].

Oral immunization generally is less efficient than parenteral vaccination in the induction of protective immune responses. For example, oral administration of HPV-11 VLPs without adjuvant was found to elicit relatively low titers of antigen-specific antibodies in serum that nevertheless demonstrated potent neutralizing activity [118]. Subsequently, if was found that co-administration of VLPs with a potent mucosal adjuvant (e.g., LT(R192G) or CpG DNA) can boost VLP oral immunogenicity to potentially useful levels [50]. In that study, VLPs of HPV-16 or HPV-18 were co-administered orally with or without *E. coli* LT(R192G) or CpG DNA. Adjuvant use was associated with the induction of high titer anti-VLP responses in serum and genital mucosal secretions. Use of LT(R192G) was also associated with the induction of Th1/Th2 antibody isotypes (i.e., IgG1 and IgG2a), whereas animals immunized with CpG DNA had a more Th1-like response [50].

Edible vaccines

Expression of antigens in transgenic plants is an attractive alternative to the use of traditional methods of vaccine production and delivery. Vaccines grown locally are economically attractive, and could facilitate the implementation of mass immunization programs.

Mammalian genes generally do not express well in plants as codon usage differs significantly between mammalian and plant systems. In addition, mammalian genes often contain regulatory sequences that can inhibit gene expression (e.g., RNA inhibitory sequences). The effects of altered codon usage [80,151] and RNA inhibitory elements [39] in the regulation of papillomavirus late gene expression have been described recently.

Hepatitis B Virus surface antigen (HbsAg) [73] and *E. coli* heat-labile enterotoxin B (LT-B) [84] have been expressed in potato, and immunogenicity has been established in mice. Norwalk Virus capsid has also been expressed in potato and found to be immunogenic in mice [83]. This material has also been found to be safe, well-tolerated and immunogenic in humans [134]. Papillomavirus L1 sequences have been synthesized for optimization of expression in plant-based systems [80] [147]. HPV capsids can assemble in potato, and transgenic potato expressing HPV-11 L1 has been found to be immunogenic in mice [147]. These encouraging results support the concept of edible vaccines for HPV immunoprophylaxis.

Immunization therapy

Prophylactic vaccination may well prove to be efficacious in preventing anogenital HPV infection; nevertheless, the incidence of anogenital HPV disease may remain high for some extended period of time due to difficulties inherent in the implementation of strategies for mass immunization, particularly in resource-poor regions. Thus, the need for better methods of treatment of established disease is

Table 1

Human autogenous vaccine studies (adapted from Abcarian and Sharon [3]

Author(s)	Cured/Total	Success rate (%)
Biberstein (1944) [12]	48/56	86
Powell et al. (1970) [111]	23/24	96
Nel and Fourie (1973) [104]	8/10	80
Ablin and Curtis (1974) [5]	1/1	100
Abcarian and Sharon (1982) [3]	190/200	95

likely to continue in the foreseeable future. The concept of "immunization therapy" as a therapeutic modality for established disease was first described by Biberstein, who demonstrated a therapeutic effect by vaccinating infected individuals with homogenates prepared from their own lesions (i.e., autogenous vaccination) [12]. Autogenous vaccination continued in use for therapy of recalcitrant/recurrent disease for approximately the next 50 years [2–4,111], but was abandoned when suspicions arose concerning HPV oncogenic potential. Nevertheless, numerous reports in the literature pre-dating the 1980s verify the successful application of this therapeutic modality (Table 1), and thus encourage greater efforts to elucidate mechanisms of viral clearance. Although results from such studies suggest that autogenous vaccine therapy may be efficacious, this technique is rarely used due to the need for relatively large amounts of infected tissue, which must be obtained from the patient's own lesions. Furthermore, vaccine preparation on an individual basis is relatively impractical.

An interesting finding from studies of autogenous vaccination suggests the possibility that antigenic type-specificity of virions may play a role in wart resolution. For example, in a study of autogenous vaccination in anal wart patients, Abcarian and Sharon noted that 45 of the 190 total respondents to this treatment modality also had warts at other anatomic sites. Of these, 40 experienced complete resolution of other mucosal warts. However, no effect was observed on cutaneous warts, when present [3], suggesting the possibility that responses to autogenous vaccination may be restricted by genotype.

Alternative strategies for immunotherapy

VLPs have been shown to enter the Major Histocompatibility Complex Class I (MHC Class I) antigen processing pathway in a phenomenon known as "cross-presentation" [9]. Thus, they can deliver foreign peptides into the class I pathway for induction of cytotoxic T lymphocyte (CTL) responses. Muller et al. have shown, for example, that regions of HPV-16 E7 can be fused with c-terminally truncated L1 to form chimeric VLPs (CVLPs), which can be used to elicit E7-specific CTL responses [64,95,120]. E7 is an attractive target for immunotherapy because it continues to be expressed in malignant disease, and thus can function as a tumor antigen. In immunization studies, L1/E7 CVLPs were found to activate CTLs *in vitro*, to inhibit the growth of E7-expressing tumors in mice, and to induce HLA-restricted T cells in humans following vaccination *in vitro* [64,95,120].

An alternative CVLP has been described in which nonfused L1 is co-expressed with a fusion protein consisting of the L2 minor capsid protein fused with an antigen of interest. Thus, L1/L2-E7 CVLPs have been generated and evaluated for the ability to elicit E7-specific responses [54]. Mice immunized with HPV-16 L1/L2-E7 CVLPs were protected from challenge with a tumor cell line expressing E7. This effect was observed in MHC class II knockout mice, but not in β2-microglobulin or perforin knockout mice, suggesting that this immunotherapeutic strategy was successful for the induction of E7-specific MHC Class I-restricted CTL [54]. TA-HPV (therapeutic antigen-human papillomavirus) is a vaccinia virus-based immunogen for potential treatment of cervical epithelial malignancy, and is potentially capable of eliciting cytotoxic T lymphocyte (CTL) responses against the E6 and E7 proteins of HPV types 16 and 18. It has been shown to be immunogenic in mice, and to induce HPV-16 E7-specific CTL [23]. It has also been evaluated in patients with late-stage cervical cancer [21]. In that study, three of eight patients that received immunization with TA-HPV developed HPV-specific antibody responses, and one of three evaluable patients demonstrated evidence of HPV-specific CTL activity [21]. Evaluations of this immunogen in a prime/boost strategy involving recombinant L2/E7 fusion proteins have been initiated [72].

E6 and E7 proteins are also attractive targets for immunotherapy of anogenital warts. A novel approach to immunotherapy of genital warts (TA-GW, therapeutic antigen-genital warts), consisting of an HPV-6 L2-E7 fusion protein, has been evaluated in humans and found to be safe, well tolerated and immunogenic, to induce antigen-specific serum IgG responses [79,135]. In that study, antigen-specific lymphoproliferative responses were also detected, suggesting the utility of this approach for generating E7-specific cellular responses.

A vaccine comprising the HPV16 L2, E6 and E7 as a single fusion protein (TA-CIN, therapeutic antigen-cervical intraepithelial neoplasia), has been evaluated in mice and found to elicit antigen-specific CTL, T-helper cells and antibodies [138]. Responses were also found to prevent the outgrowth of an HPV-16 tumor cell line.

The ability of HPV-16 E7-specific CTL to inhibit the outgrowth of tumor cell lines expressing this protein in mice [54,120] supports the potential feasibility of using this strategy to induce and/or enhance cellular immune responses in individuals with persistent disease. However, evidence of disease persistence in individuals with detectable E7-specific CTL responses suggests the possibility that alternative CTL specificities may be required for eradication of established disease [102].

Conclusions

Recognition of the essential role played by certain human papillomaviruses in the pathogenesis of cervical cancer and other epithelial malignancies has stimulated efforts to develop vaccines capable of preventing such diseases. If this becomes possible, new strategies will be needed to facilitate mass immunizations. Also remaining is an urgent need for additional information concerning the nature of effective immune responses to infection, to guide the development of improved therapies for established disease.

References

1. Aaltonen LM, Auvinen E, Dillner J, Lehtinen M, Paavonen J, Rihkanen H, Vaheri A. Poor antibody response against human papillomavirus in adult-onset laryngeal papillomatosis. J Med Microbiol 2001; 50: 468–71.
2. Abcarian H, Sharon N. The effectiveness of immunotherapy in the treatment of anal condyloma acuminatum. J Surg Res 1977; 22: 231–236.
3. Abcarian H, Sharon N. Long-term effectiveness of the immunotherapy of anal condyloma acuminatum. Dis Colon & Rectum 1982; 25: 648–651.
4. Abcarian H, Smith D, Sharon N. The immunotherapy of anal condyloma acuminatum. Dis Colon & Rectum 1976; 19: 237–244.
5. Ablin RJ, Curtis WW. Immunotherapeutic treatment of condyloma acuminata. Gynecol Oncol 1974; 2: 446–450.
6. al-Saleh W, Giannini SL, Jacobs N, Moutschen M, Doyen J, Boniver J, Delvenne P. Correlation of T-helper secretory differentiation and types of antigen-presenting cells in squamous intraepithelial lesions of the uterine cervix. J Pathol 1998; 184: 283–290.
7. Amerongen HM, Weltzin R, Farnet CM, Michetti P, Haseltine WA, Neutra MR. Transepithelial transport of HIV-1 by intestinal M cells: a mechanism for transmission of AIDS. J Acquired Immune Def Syndromes 1991; 4: 760–765.
8. Anisimova E, Bartak P, Vlcek D, Hirsch I, Brichacek B, Vonka V, Presence and type specificity of papillomavirus antibodies demonstrable by immunoelectron microscopy tests in samples from patients with warts. J Gen Virol 1990; 71: 419–422.
9. Bachmann MF, Lutz MB, Layton GT, Harris SJ, Fehr T, Rescigno M, Ricciardi-Castagnoli P. Dendritic cells process exogenous viral proteins and virus-like particles for class I presentation to CD8+ cytotoxic T lymphocytes. Eur J Immunol 1996; 26: 2595–600.
10. Ball JM, Hardy ME, Atmar RL, Conner ME, Estes MK. Oral immunization with recombinant norwalk virus-like particles induces a systemic and mucosal immune response in mice. J Virol 1998; 72: 1345–1353.
11. Balmelli C, Roden R, Potts A, Schiller J, De Grandi P, Nardelli-Haefliger D. Nasal immunization of mice with human papillomavirus type 16 virus-like particles elicits neutralizing antibodies in mucosal secretions. J Virol 1998; 72: 8220–8229.
12. Biberstein H. Immunization therapy of warts. Arch Dermatol Syphilol 1944; 50: 12–22.
13. Bleul C, Muller M, Frank R, Gausepohl H, Koldovsky U, Mgaya HN, Luande J, Pawlita M, ter Meulen J, Viscidi R. et al. Human papillomavirus type 18 E6 and E7 antibodies in human sera: increased anti-E7 prevalence in cervical cancer patients. J Clin Microbiol 1991; 29: 1579–1588.
14. Bonnez W, Da Rin C, Rose RC, Reichman RC. Use of human papillomavirus type 11 virions in an ELISA to detect specific antibodies in humans with condylomata acuminata. J Gen Virol 1991; 72: 1343–1347.

15. Bonnez W, Da Rin C, Rose RC, Tyring SK, Reichman RC. Evolution of the antibody response to human papillomavirus type 11 (HPV-11) in patients with condyloma acuminatum according to treatment response. J Med Virol 1993; 39: 340–344.

16. Bonnez W, DaRin C, Borkhuis C, de Mesy Jensen K, Reichman RC, Rose RC. Isolation and propagation of human papillomavirus type 16 in human xenografts implanted in the severe combined immunodeficiency mouse. J Virol 1998; 72(6): 5256–5261.

17. Bonnez W, Kashima HK, Leventhal B, Mounts P, Rose RC, Reichman RC, Shah KV. Antibody response to human papillomavirus (HPV) type 11 in children with juvenile-onset recurrent respiratory papillomatosis (RRP). Virology 1992; 188: 384–387.

18. Bonnez W, Rose RC, Reichman RC. Antibody-mediated neutralization of human papillomavirus type 11 (HPV-11) infection in the nude mouse: detection of HPV-11 mRNAs. J Infect Dis 1992; 165: 376–380.

19. Bontkes HJ, de Gruijl TD, Bijl A, Verheijen RH, Meijer CJ, Scheper RJ, Stern PL, Burns JE, Maitland NJ, Walboomers JM. Human papillomavirus type 16 E2-specific T-helper lymphocyte responses in patients with cervical intraepithelial neoplasia. J Gen Virol 1999; 80: 2453–2459.

20. Bontkes HJ, de Gruijl TD, Walboomers JM, Schiller JT, Dillner J, Helmerhorst TJ, Verheijen RH, Scheper RJ, Meijer CJ. Immune responses against human papillomavirus (HPV) type 16 virus-like particles in a cohort study of women with cervical intraepithelial neoplasia. II. Systemic but not local IgA responses correlate with clearance of HPV-16. J Gen Virol 1999; 80: 409–417.

21. Borysiewicz LK, Fiander A, Nimako M, Man S, Wilkinson GW, Westmoreland D, Evans AS, Adams M, Stacey SN, Boursnell ME, Rutherford E, Hickling JK, Inglis SC. A recombinant vaccinia virus encoding human papillomavirus types 16 and 18, E6 and E7 proteins as immunotherapy for cervical cancer. Lancet 1996; 347: 1523–1527.

22. Bosch FX, Lorincz A, Munoz N, Meijer CJ, Shah KV. The causal relation between human papillomavirus and cervical cancer. J Clin Pathol 2002; 55: 244–265.

23. Boursnell ME, Rutherford E, Hickling JK, Rollinson EA, Munro AJ, Rolley N, McLean CS, Borysiewicz LK, Vousden K, Inglis SC. Construction and characterisation of a recombinant vaccinia virus expressing human papillomavirus proteins for immuno-therapy of cervical cancer. Vaccine 1996; 14: 1485–1494.

24. Breitburd F, Kirnbauer R, Hubbert NL, Nonnenmacher B, Trin-Dinh-Desmarquet C, Orth G, Schiller JT, Lowy DR. Immunization with viruslike particles from cottontail rabbit papillomavirus (CRPV) can protect against experimental CRPV infection. J Virol 1995; 69: 3959–3963.

25. Buller CR, Moxley RA. Natural infection of porcine ileal dome M cells with rotavirus and enteric adenovirus. Vet Pathol 1988; 25: 516–517.

26. Campo MS, Grindlay GJ, O'Neil BW, Chandrachud LM, McGarvie GM, Jarrett WFH. Prophylactic and therapeutic vaccination against a mucosal papillomavirus. J Gen Virol 1993; 74: 945–953.

27. Cardenas-Freytag L, Cheng E, Mayeux P, Domer JE, Clements JD. Effectiveness of a vaccine composed of heat-killed Candida albicans and a novel mucosal adjuvant, LT(R192G), against systemic candidiasis. Infect Immunol 1999; 67: 826–833.

28. Cardenas-Freytag L, Cheng E, Mirza A. New approaches to mucosal immunization. Adv Exper Med Biol 1999; 473: 319–337.

29. Carter JJ, Hagensee M, Taflin MC, Lee SK, Koutsky LA, Galloway DA. HPV-1 capsids

expressed *in vitro* detect human serum antibodies associated with foot warts. Virology 1993; 195: 456–462.

30. Carter JJ, Koutsky LA, Hughes JP, Lee SK, Kuypers J, Kiviat N, Galloway DA. Comparison of human papillomavirus types 16, 18, and 6 capsid antibody responses following incident infection. J Infect Dis 2000; 181: 1911–1919.

31. Carter JJ, Koutsky LA, Wipf GC, Christensen ND, Lee SK, Kuypers J, Kiviat N, Galloway DA. The natural history of human papillomavirus type 16 capsid antibodies among a cohort of university women. J Infect Dis 1996; 174: 927–936.

32. Chen LP, Thomas EK, Hu SL, Hellstrom I, Hellstrom KE. Human papillomavirus type 16 nucleoprotein E7 is a tumor rejection antigen. Proc Natl Acad Sci USA 1991; 88: 110–114.

33. Choi AH, Basu M, McNeal MM, Flint J, VanCott JL, Clements JD, Ward RL. Functional mapping of protective domains and epitopes in the rotavirus VP6 protein [In Process Citation]. J Virol 2000; 74: 11574–11580.

34. Christensen ND, Hopfl R, DiAngelo SL, Cladel NM, Patrick SD, Welsh PA, Budgeon LR, Reed CA, Kreider JW. Assembled baculovirus-expressed human papillomavirus type 11 L1 capsid protein virus-like particles are recognized by neutralizing monoclonal antibodies and induce high titres of neutralizing antibodies. J Gen Virol 1994; 75: 2271–2276.

35. Christensen ND, Kreider JW. Antibody-mediated neutralization *in vivo* of infectious papillomaviruses. J Virol 1990; 64: 3151–3156.

36. Christensen ND, Kreider JW, Cladel NM, Patrick SD, Welsh PA. Monoclonal antibody-mediated neutralization of infectious human papillomavirus type 11. J Virol 1990; 64: 5678–5681.

37. Christensen ND, Kreider JW, Shah KV, Rando RF. Detection of human serum antibodies that neutralize infectious human papillomavirus type 11 virions. J Gen Virol 1992; 73: 1261–1267.

38. Coleman N, Birley HD, Renton AM, Hanna NF, Ryait BK, Byrne M, Taylor-Robinson D, Stanley MA. Immunological events in regressing genital warts. Am J Clin Pathol 1994; 102: 768–774.

39. Collier B, Oberg D, Zhao X, Schwartz S. Specific inactivation of inhibitory sequences in the 5' end of the human papillomavirus type 16 L1 open reading frame results in production of high levels of L1 protein in human epithelial cells. J Virol 2002; 76: 2739–2752.

40. Davidson EJ, Brown MD, Burt DJ, Parish JL, Gaston K, Kitchener HC, Stacey SN, Stern PL. Human T cell responses to HPV 16 E2 generated with monocyte-derived dendritic cells. Int J Cancer 2001; 94: 807–812.

41. de Aizpurua HJ, Russell-Jones GJ. Oral vaccination. Identification of classes of proteins that provoke an immune response upon oral feeding. J Exper Med 1988; 167: 440–451.

42. de Gruijl TD, Bontkes HJ, Walboomers JM, Coursaget P, Stukart MJ, Dupuy C, Kueter E, Verheijen RH, Helmerhorst TJ, Duggan-Keen MF, Stern PL, Meijer CJ, Scheper RJ. Immune responses against human papillomavirus (HPV) type 16 virus-like particles in a cohort study of women with cervical intraepithelial neoplasia. I. Differential T-helper and IgG responses in relation to HPV infection and disease outcome. J Gen Virol 1999; 80: 399–408.

43. de Gruijl TD, Bontkes HJ, Walboomers JM, Stukart MJ, Doekhie FS, Remmink AJ,

Helmerhorst TJ, Verheijen RH, Duggan-Keen MF, Stern PL, Meijer CJ, Scheper RJ. Differential T helper cell responses to human papillomavirus type 16 E7 related to viral clearance or persistence in patients with cervical neoplasia: a longitudinal study. Cancer Res 1998; 58: 1700–1706.

44.   de Gruijl TD, Bontkes HJ, Walboomers JMM, Schiller JT, Stukart MJ, Groot BS, Chabaud MMR, Remmink AJ, Verheijen RHM, Helmerhorst TJM, Meijer CJLM, Scheper RJ. Immunoglobulin g responses against human papillomavirus type 16 virus-like particles in a prospective nonintervention cohort study of women with cervical intraepithelial neoplasia. J Nat Cancer Inst 1997; 89: 630–638.

45.   Dickinson BL, Clements JD. Dissociation of *Escherichia coli* heat-labile enterotoxin adjuvanticity from ADP-ribosyltransferase activity. Infection Immunity 1995; 63: 1617–1623.

46.   Dowdle WR, Featherstone DA, Birmingham ME, Hull HF, Aylward RB. Poliomyelitis eradication. Virus Res 1999; 62: 185–192.

47.   Dupuy C, Buzoni-Gatel D, Touze A, Bout D, Coursaget P. Nasal immunization of mice with human papillomavirus type 16 (HPV-16) virus-like particles or with the HPV-16 L1 gene elicits specific cytotoxic T lymphocytes in vaginal draining lymph nodes. J Virol 1999; 73: 9063–9071.

48.   Dvoretzky I, Shober R, Chattopadhyay S, Lowy D. A quantitative *in vitro* focus assay for bovine papilloma virus. Virology 1980. 103: 369–375.

49.   Evans TG, Bonnez W, Rose RC, Koenig S, Demeter L, Suzich JA, O'Brien D, Campbell M, White WI, Balsley J, Reichman RC. A Phase 1 study of a recombinant viruslike particle vaccine against human papillomavirus type 11 in healthy adult volunteers. J Infect Dis 2001; 183: 1485–1493.

50.   Gerber S, Lane C, Brown DM, Lord E, DiLorenzo M, Clements JD, Rybicki E, Williamson AL, Rose RC. Human papillomavirus virus-like particles are efficient oral immunogens when coadministered with *Escherichia coli* heat-labile enterotoxin mutant R192G or CpG DNA. J Virol 2001; 75: 4752–4760.

51.   Giroglou T, Sapp M, Lane C, Fligge C, Christensen ND, Streeck RE, Rose RC. Immuno-logical analyses of human papillomavirus capsids. Vaccine 2001; 19: 1783–1793.

52.   Gissmann L, Pfister H, Zur Hausen H. Human papilloma viruses (HPV): characteri-zation of four different isolates. Virology 1977; 76: 569–580.

53.   GLOBOCAN 2001; GLOBOCAN Cancer Incidence, Mortality and Prevalence World-wide, Version 1.0. IARC CancerBase No. 5. IARC Press, Lyon, 2001.

54.   Greenstone HL, Nieland JD, de Visser KE, De Bruijn ML, Kirnbauer R, Roden RB, Lowy DR, Kast WM, Schiller JT. Chimeric papillomavirus virus-like particles elicit antitumor immunity against the E7 oncoprotein in an HPV16 tumor model. Proc Natl Acad Sci USA 1998; 95: 1800–1805.

55.   Greer CE, Petracca R, Buonamassa DT, Di Tommaso A, Gervase B, Reeve RL, Ugozzoli M, Van Nest G, De Magistris MT, Bensi G. The comparison of the effect of LTR72 and MF59 adjuvants on mouse humoral response to intranasal immunisation with human papillomavirus type 6b (HPV-6b) virus-like particles. Vaccine 2000; 19: 1008–1012.

56.   Guillobel HC, Carinhanha JI, Cardenas L, Clements JD, de Almeida DF, Ferreira LC. Adjuvant activity of a nontoxic mutant of *Escherichia coli* heat-labile enterotoxin on systemic and mucosal immune responses elicited against a heterologous antigen carried

by a live *Salmonella enterica* serovar *Typhimurium* vaccine strain. Infect Immun 2000; 68: 4349–4353.

57. Hagensee ME, Koutsky LA, Lee SK, Grubert T, Kuypers J, Kiviat NB, Galloway DA. Detection of cervical antibodies to human papillomavirus type 16 (HPV-16) capsid antigens in relation to detection of HPV-16 DNA and cervical lesions. J Infect Dis 2000; 181: 1234–1239.

58. Hagensee ME, Yaegashi N, Galloway DA. Self-assembly of human papillomavirus type 1 capsids by expression of the L1 protein alone or by coexpression of the L1 and L2 capsid proteins. J Virol 1993; 67: 315–322.

59. Halpert R, Fruchter RG, Sedlis A, Butt K, Boyce JG, Sillman FH. Human papillomavirus and lower genital neoplasia in renal transplant patients. Obstet Gynecol 1986. 68: 251–258.

60. Harro CD, Pang YY, Roden RB, Hildesheim A, Wang Z, Reynolds MJ, Mast TC, Robinson R, Murphy BR, Karron RA, Dillner J, Schiller JT, Lowy DR. Safety and immunogenicity trial in adult volunteers of a human papillomavirus 16 L1 virus-like particle vaccine. J Natl Cancer Inst 2001; 93: 284–292.

61. Hemmi H, Takeuchi O, Kawai T, Kaisho T, Sato S, Sanjo H, Matsumoto M, Hoshino K, Wagner H, Takeda K, Akira S. A Toll-like receptor recognizes bacterial DNA.[In Process Citation]. Nature 2000; 408: 740–745.

62. Inman LR, Cantey JR. Specific adherence of *Escherichia coli* (strain RDEC-1) to membranous (M) cells of the Peyer's patch in *Escherichia coli* diarrhea in the rabbit. J Clin Invest 1983. 71: 1–8.

63. Jarrett WF, Smith KT, O'Neil BW, Gaukroger JM, Chandrachud LM, Grindlay GJ, McGarvie GM, Campo MS. Studies on vaccination against papillomaviruses: prophylactic and therapeutic vaccination with recombinant structural proteins. Virology 1991; 184: 33–42.

64. Kaufmann AM, Nieland J, Schinz M, Nonn M, Gabelsberger J, Meissner H, Muller RT, Jochmus I, Gissmann L, Schneider A, Durst M. HPV16 L1E7 chimeric virus-like particles induce specific HLA-restricted T cells in humans after *in vitro* vaccination. Int J Cancer 2001; 92: 285–93.

65. Keren DF, McDonald RA, Wassef JS, Armstrong LR, Brown JE. The enteric immune response to shigella antigens. [Review] [21 refs]. Curr Top Microbiol Immunol 1989; 146: 213–223.

66. Kienzler JL, Lemoine MT, Orth G, Jibard N, Blanc D, Laurent R, Agache P. Humoral and cell-mediated immunity to human papillomavirus type 1 (HPV-1) in human warts. Br J Dermatol 1983; 108: 665–672.

67. Kirnbauer R. 1996; Papillomavirus-like particles for serology and vaccine development. Intervirology 39: 54–61.

68. Kirnbauer R, Booy F, Cheng N, Lowy DR, Schiller JT. Papillomavirus L1 major capsid protein self-assembles into virus-like particles that are highly immunogenic. Proc Natl Acad Sci USA 1992; 89: 12180–12184.

69. Kirnbauer R, Chandrachud LM, O'Neil BW, Wagner ER, Grindlay GJ, Armstrong A, McGarvie GM, Schiller JT, Lowy DR, Campo MS. Virus-like particles of bovine papillomavirus type 4 in prophylactic and therapeutic immunization. Virology 1996; 219: 37–44.

70. Kirnbauer R, Hubbert NL, Wheeler CM, Becker TM, Lowy DR, Schiller JT. A virus-like

particle enzyme-linked immunosorbent assay detects serum antibodies in a majority of women infected with human papillomavirus type 16 [see comments]. J Nat Cancer Inst 1994; 86: 494–499.

71. Kirnbauer R, Taub J, Greenstone H, Roden R, Durst M, Gissmann L, Lowy DR, Schiller JT. Efficient self-assembly of human papillomavirus type 16 L1 and L1-L2 into virus-like particles. J Virol 1993; 67: 6929–6936.

72. Knutson KL. Technology evaluation: T-cell activator, Xenova. Curr Opin Mol Ther 2001; 3: 585–588.

73. Kong Q, Richter L, Yang YF, Arntzen CJ, Mason HS, Thanavala Y. Oral immunization with hepatitis B surface antigen expressed in transgenic plants. Proc Natl Acad Sci USA 2001; 98: 11539–11544.

74. Konya J, Dillner J. Immunity to oncogenic human papillomaviruses. Adv Cancer Res 2001; 82: 205–238.

75. Konya J, Eklund C, af Geijersstam V, Yuan F, Stuber G, Dillner J. Identification of a cytotoxic T-lymphocyte epitope in the human papillomavirus type 16 E2 protein. J Gen Virol 1997; 78: 2615–2620.

76. Kreider JW, Howett MK. Leure-Dupree AE, Zaino RJ, Weber JA. Laboratory production {in vivo} of infectious human papillomavirus type 11. J Virol 1987; 61: 590–593.

77. Kreider JW, Howett MK, Wolfe SA, Bartlett GL, Zaino RJ, Sedlacek T, Mortel R. Morphological transformation in vivo of human uterine cervix with papillomavirus from condylomata acuminata. Nature 1985; 317: 639–641.

78. Krieg AM, Yi AK, Matson S, Waldschmidt TJ, Bishop GA, Teasdale R, Koretzky GA, Klinman DM. CpG motifs in bacterial DNA trigger direct B-cell activation. Nature 1995; 374: 546–549.

79. Lacey CJ, Thompson HS, Monteiro EF, O'Neill T, Davies ML, Holding FP, Fallon RE, Roberts JS. Phase IIa safety and immunogenicity of a therapeutic vaccine, TA-GW, in persons with genital warts. J Infect Dis 1999; 179: 612–618.

80. Leder C, Kleinschmidt JA, Wiethe C, Muller M. Enhancement of capsid gene expression: preparing the human papillomavirus type 16 major structural gene L1 for DNA vaccination purposes. J Virol 2001; 75: 9201–9209.

81. Lehtinen M, Hibma MH, Stellato G, Kuoppala T, Paavonen J. Human T helper cell epitopes overlap B cell and putative cytotoxic T cell epitopes in the E2 protein of human papillomavirus type 16. Biochem Biophys Res Commun 1995; 209: 541–546.

82. Marais DJ, Best JM, Rose RC, Keating P, Soeters R, Denny L, Dehaeck CM, Nevin J, Kay P, Passmore JA, Williamson AL. Oral antibodies to human papillomavirus type 16 in women with cervical neoplasia. J Med Virol 2001; 65: 149–154.

83. Mason HS, Ball JM, Shi JJ, Jiang X, Estes MK, Arntzen CJ. Expression of Norwalk virus capsid protein in transgenic tobacco and potato and its oral immunogenicity in mice. Proc Natl Acad Sci USA 1996; 93: 5335–5340.

84. Mason HS, Haq TA, Clements JD, Arntzen CJ. Edible vaccine protects mice against escherichia coli heat-labile enterotoxin (lt)—potatoes expressing a synthetic lt-b gene. Vaccine 1998; 16: 1336–1343.

85. McCluskie MJ, Davis HL. CpG DNA as mucosal adjuvant. Vaccine 1999; 18: 231–237.

86. McGhee JR, Czerkinsky C, Mestecky J. Mucosal vaccines: an overview. In: PL Ogra, J. Mestecky, ME Lamm, W. Strober, J. Bienenstock, JR McGhee (Eds), Mucosal Immunology, 2nd Edn. Academic Press, London, 1999, pp. 741–757.

87. Meneguzzi G, Cerni C, Kieny MP, Lathe R. Immunization against human papillomavirus type 16 tumor cells with recombinant vaccinia viruses expressing E6 and E7. Virology 1991; 181: 62–69.

88. Meneguzzi G, Kieny MP, Lecocq JP, Chambon P, Cuzin F, Lathe R. Vaccinia recombinants expressing early bovine papilloma virus (BPV1) proteins: retardation of BPV1 tumour development. Vaccine 1990; 8: 199–204.

89. Mestecky J. The common mucosal immune system and current strategies for induction of immune responses in external secretions. J Clin Immunol 1987; 7: 265–276.

90. Misra N, Murtaza A, Walker B, Narayan NP, Misra RS, Ramesh V, Singh S, Colston MJ, Nath I. Cytokine profile of circulating T cells of leprosy patients reflects both indiscriminate and polarized T-helper subsets: T-helper phenotype is stable and uninfluenced by related antigens of Mycobacterium leprae. Immunology 1995; 86: 97–103.

91. Moldoveanu Z, Love-Homan L, Huang WQ, Krieg AM. CpG DNA, a novel immune enhancer for systemic and mucosal immunization with influenza virus. Vaccine 1998; 16: 1216–1224.

92. Morison WL. *In vitro* assay of immunity to human wart antigen. Br J Dermatol 1975. 93: 545–552.

93. Mosmann TR, Cherwinski H, Bond MW, Giedlin MA, Coffman RL. Two types of murine helper T cell clone. I. Definition according to profiles of lymphokine activities and secreted proteins. J Immunol 1986. 136: 2348–2357.

94. Muller M, Gissmann L, Cristiano RJ, Sun XY, Frazer IH, Jenson AB, Alonso A, Zentgraf H, Zhou J. Papillomavirus capsid binding and uptake by cells from different tissues and species. J Virol 1995; 69: 948–954.

95. Muller M, Zhou J, Reed TD, Rittmuller C, Burger A, Gabelsberger J, Braspenning J, Gissmann L. Chimeric papillomavirus-like particles. Virology 1997; 234: 93–111.

96. Munos N. Disease-burden related to cancer induced by human viruses and *H. pylori*. World Health Organization (WHO) Vaccine Research and Development: Report of the Technical Review Group Meeting, 9–10 June, 1997/

97. Murphy BR. Mucosal immunity to viruses. In: PL Ogra, J. Mestecky, ME Lamm, W. Strober, J. Bienenstock, JR McGhee (Eds), Mucosal Immunology. Academic Press, San Diego, 1999, pp. 695–707.

98. Nabors GS, Afonso LC, Farrell JP, Scott P. Switch from a type 2 to a type 1 T helper cell response and cure of established Leishmania major infection in mice is induced by combined therapy with interleukin 12 and Pentostam. Proc Natl Acad Sci USA 1995; 92: 3142–3146.

99. Nakagawa M, Stites DP, Farhat S, Judd A, Moscicki AB, Canchola AJ, Hilton JF, Palefsky JM. T-cell proliferative response to human papillomavirus type 16 peptides: relationship to cervical intraepithelial neoplasia. Clin Diagn Lab Immunol. 1996; 3: 205–210.

100. Nakagawa M, Stites DP, Farhat S, Sisler JR, Moss B, Kong F, Moscicki AB, Palefsky JM. Cytotoxic T lymphocyte responses to E6 and E7 proteins of human papillomavirus type 16: relationship to cervical intraepithelial neoplasia. J Infect Dis 1997; 175: 927–931.

101. Nakagawa M, Stites DP, Palefsky JM, Kneass Z, Moscicki AB. CD4-positive and CD8-positive cytotoxic T lymphocytes contribute to human papillomavirus type 16 E6 and E7 responses. Clin Diagn Lab Immunol 1999; 6: 494–498.

102. Nakagawa M, Stites DP, Patel S, Farhat S, Scott M, Hills NK, Palefsky JM, Moscicki AB.

Persistence of human papillomavirus type 16 infection is associated with lack of cytotoxic T lymphocyte response to the E6 antigens. J Infect Dis 2000; 182: 595–598.

103. Nardelli-Haefliger D, Roden R, Balmelli C, Potts A, Schiller J, De Grandi P. Mucosal but not parenteral immunization with purified human papillomavirus type 16 virus-like particles induces neutralizing titers of antibodies throughout the estrous cycle of mice. J Virol 1999; 73: 9609–9613.

104. Nel WS, Fourie ED. Immunotherapy and 5% topical 5-fluoro-uracil ointment in the treatment of condylomata acuminata. S Afr Med J 1973. 47: 45–49.

105. Neutra MR, Pringault E, Kraehenbuhl JP. Antigen sampling across epithelial barriers and induction of mucosal immune responses. [Review] [148 refs]. Annu Rev Immunol 1996; 14: 275–300.

106. Oneal CM, Crawford SE, Estes MK, Conner ME. Rotavirus virus-like particles administered mucosally induce protective immunity. J Virol 1997; 71: 8707–8717.

107. Orth G, Favre M, Croissant O. Characterization of a new type of human papillomavirus that causes skin warts. J Virol 1977. 24: 108–120.

108. Pao CC, Lin CY, Yao DS, Tseng CJ. Differential expression of cytokine genes in cervical cancer tissues. Biochem Biophys Res Commun 1995; 214: 1146–1151.

109. Pfister H, zur Hausen H. Characterization of proteins of human papilloma viruses (HPV) and antibody response to HPV 1. Med Microbiol Immunol (Berl) 1978. 166: 13–19.

110. Pirmez C, Yamamura M, Uyemura K, Paes-Oliveira M, Conceicao-Silva F, Modlin RL. Cytokine patterns in the pathogenesis of human leishmaniasis. J Clin Invest 1993; 91: 1390–1395.

111. Powell Jr LC, Pollard M, Jinkins Sr JL. Treatment of condyloma acuminata by auto-genous vaccine. Southern Med J 1970. 63: 202–205.

112. Rocha-Zavaleta L, Barrios T, Garcia-Carranca A, Valdespino V, Cruz-Talonia F. Cervical secretory immunoglobulin A to human papillomavirus type 16 (HPV16) from HPV16-infected women inhibit HPV16 virus-like particles-induced hemagglutination of mouse red blood cells. FEMS Immunol Med Microbiol 2001; 31: 47–51.

113. Rocha-Zavaleta L, Jordan D, Pepper S, Corbitt G, Clarke F, Maitland NJ, Sanders CM, Arrand JR, Stern PL, Stacey SN. Differences in serological IgA responses to recombinant baculovirus-derived human papillomavirus E2 protein in the natural history of cervical neoplasia. Br J Cancer 1997; 75: 1144–1150.

114. Roden RB, Greenstone HL, Kirnbauer R, Booy FP, Jessie J, Lowy DR, Schiller JT. In vitro generation and type-specific neutralization of a human papillomavirus type 16 virion pseudotype. J Virol 1996; 70: 5875–5883.

115. Roden RB, Hubbert NL, Kirnbauer R, Christensen ND, Lowy DR, Schiller JT. Assessment of the serological relatedness of genital human papillomaviruses by hemagglutination inhibition. J Virol 1996; 70: 3298–3301.

116. Rose RC, Bonnez W, Da Rin C, McCance DJ, Reichman RC. Serological differentiation of human papillomavirus types 11, 16 and 18 using recombinant virus-like particles. J Gen Virol 1994; 75: 2445–2449.

117. Rose RC, Bonnez W, Reichman RC, Garcea RL. Expression of human papillomavirus type 11 L1 protein in insect cells: in vivo and in vitro assembly of viruslike particles. J Virol

1993; 67: 1936–1944.

118. Rose RC, Lane C, Wilson S, Suzich JA, Rybicki E, Williamson AL. Oral vaccination of mice with human papillomavirus virus-like particles induces systemic virus-neutralizing antibodies. Vaccine 1999; 17: 2129–2135.

119. Rose RC, Reichman RC, Bonnez W. Human papillomavirus (HPV) type 11 recombinant virus-like particles induce the formation of neutralizing antibodies and detect HPV-specific antibodies in human sera. J Gen Virol 1994; 75: 2075–2079.

120. Schafer K, Muller M, Faath S, Henn A, Osen W, Zentgraf H, Benner A, Gissmann L, Jochmus I. Immune response to human papillomavirus 16 L1E7 chimeric virus-like particles: induction of cytotoxic T cells and specific tumor protection. Int J Cancer 1999; 81: 881–888.

121. Schiller JT. Papillomavirus-like particle vaccines for cervical cancer. [Review] [33 refs]. Molec Med Today 1999; 5: 209–215.

122. Schiller JT, Lowy DR. Papillomavirus-like particles and HPV vaccine development. Semin Cancer Biol 1996; 7: 373–382.

123. Scott M, Nakagawa M, Moscicki AB. Cell-mediated immune response to human papillomavirus infection. Clin Diagn Lab Immunol 2001; 8: 209–220.

124. Scott M, Stites DP, Moscicki AB. Th1 cytokine patterns in cervical human papillomavirus infection. Clin Diagn Lab Immunol 1999; 6: 751–755.

125. Segre D, Olson C, Hoerlein AB. Neutralization of bovine papilloma virus with serums from cattle and horses with experimental papillomas. Am J Veter Res 1955. 16: 517–520.

126. Shirodaria PV, Matthews RS. An immunofluorescence study of warts. Clin & Exper Immunol 1975; 21: 329–338.

127. Sicinski P, Rowinski J, Warchol JB, Jarzabek Z, Gut W, Szczygiel B, Bielecki K, Koch G. Poliovirus type 1 enters the human host through intestinal M cells. Gastroenterology 1990; 98: 56–58.

128. Sillman F, Stanek A, Sedlis A, Rosenthal J, Lanks KW, Buchhagen D, Nicastri A, Boyce J. The relationship between human papillomavirus and lower genital intraepithelial neoplasia in immunosuppressed women. Am J Obstet Gynecol 1984; 150: 300–308.

129. Sillman FH, Sedlis A. Anogenital papillomavirus infection and neoplasia in immuno-deficient women. Obstet Gynecol Clin North Am 1987; 14: 537–558.

130. Steele JC, Gallimore PH. Humoral assays of human sera to disrupted and nondisrupted epitopes of human papillomavirus type 1. Virology 1990; 174: 388–398.

131. Steele JC, Stankovic T, Gallimore PH. Production and characterization of human proliferative T-cell clones specific for human papillomavirus type 1 E4 protein. J Virol 1993; 67: 2799–2806.

132. Strang G, Hickling JK, McIndoe GA, Howland K, Wilkinson D, Ikeda H, Rothbard JB. Human T cell responses to human papillomavirus type 16 L1 and E6 synthetic peptides: identification of T cell determinants, HLA-DR restriction and virus type specificity. J Gen Virol 1990; 71: 423–431.

133. Suzich JA, Ghim SJ, Palmer-Hill FJ, White WI, Tamura JK, Bell JA, Newsome JA, Jenson AB, Schlegel R. Systemic immunization with papillomavirus L1 protein completely prevents the development of viral mucosal papillomas. Proc Natl Acad Sci USA 1995; 92: 11553–11557.

134. Tacket CO, Mason HS, Losonsky G, Estes MK, Levine MM, Arntzen CJ. Human immune responses to a novel norwalk virus vaccine delivered in transgenic potatoes. J Infect Dis 2000; 182: 302–305.

135. Thompson HS, Davies ML, Holding FP, Fallon RE, Mann AE, O'Neill T, Roberts JS. Phase I safety and antigenicity of TA-GW: a recombinant HPV6 L2E7 vaccine for the treatment of genital warts. Vaccine 1999; 17: 40–49.

136. Tjiong MY, Zumbach K, Schegget JT, van der Vange N, Out TA, Pawlita M, Struyk L. Antibodies against human papillomavirus type 16 and 18 E6 and E7 proteins in cervico-vaginal washings and serum of patients with cervical neoplasia. Viral Immunol 2001; 14: 415–24.

137. Unckell F, Streeck RE, Sapp M. Generation and neutralization of pseudovirions of human papillomavirus type 33. J Virol 1997; 71: 2934–2939.

138. van der Burg SH, Kwappenberg KM, O'Neill T, Brandt RM, Melief CJ, Hickling JK, Offringa R. Pre-clinical safety and efficacy of TA-CIN, a recombinant HPV16 L2E6E7 fusion protein vaccine, in homologous and heterologous prime-boost regimens. Vaccine 2001; 19: 3652–3660.

139. Viac J, Chomel JJ, Chardonnet Y, Aymard M. Incidence of antibodies to human papillomavirus type 1 in patients with cutaneous and mucosal papillomas. J Med Virol 1990; 32: 18–21.

140. Viac J, Staquet MJ, Thivolet J. [Antigenic relations between human papilloma viruses (HPV): experimental study in guinea-pigs (author's transl]. Ann Immunol (Paris) 1978; 129 C: 559–570.

141. Viscidi RP, Sun Y, Tsuzaki B, Bosch FX, Munoz N, Shah KV. Serologic response in human papillomavirus-associated invasive cervical cancer. Int J Cancer 1993; 55: 780–784.

142. Volpers C, Schirmacher P, Streeck RE, Sapp M. Assembly of the major and the minor capsid protein of human papillomavirus type 33 into virus-like particles and tubular structures in insect cells. Virology 1994; 200: 504–512.

143. von Krogh G. Warts: Immunologic factors of prognostic significance. Int J Dermatol 1979; 18: 195–204.

144. Wagner H. Bacterial CpG DNA activates immune cells to signal infectious danger. Adv Immunol 1999; 73: 329–368.

145. Wang Z, Konya J, Avall-Lundkvist E, Sapp M, Dillner J, Dillner L. Human papilloma-virus antibody responses among patients with incident cervical carcinoma. J Med Virol 1997; 52: 436–440.

146. Wang ZH, Hansson BG, Forslund O, Dillner L, Sapp M, Schiller JT, Bjerre B, Dillner J. Cervical mucus antibodies against human papillomavirus type 16, 18, and 33 capsids in relation to presence of viral DNA. J Clin Microbiol 1996; 34: 3056–3062.

147. Warzecha HM, Mason HS, Arntzen CJ, Clements JD, Lane C, Rose RC. Oral immuno-genicity of human papillomavirus virus-like particles expressed in potato. In preparation.

148. White WI, Wilson SD, Bonnez W, Rose RC, Koenig S, Suzich JA. *In vitro* infection and type-restricted antibody-mediated neutralization of authentic human papillomavirus type 16. J Virol 1998; 72: 959–964.

149. Wideroff L, Schiffman MH, Nonnenmacher B, Hubbert N, Kirnbauer R, Greer CE, Lowy D, Lorincz AT, Manos MM, Glass AG et al. Evaluation of seroreactivity to human

papillomavirus type 16 virus-like particles in an incident case-control study of cervical neoplasia. J Infect Dis 1995; 172: 1425–1430.

150. Wolf JL, Rubin DH, Finberg R, Kauffman RS, Sharpe AH, Trier JS, Fields BN. Intestinal M cells: a pathway for entry of reovirus into the host. Science 1981. 212: 471–472.

151. Zhou J, Liu WJ, Peng SW, Sun XY, Frazer I. Papillomavirus capsid protein expression level depends on the match between codon usage and tRNA availability. J Virol 1999; 73: 4972–4982.

152. Zhou J, Stenzel DJ, Sun XY, Frazer IH. Synthesis and assembly of infectious bovine papillomavirus particles *in vitro*. J Gen Virol 1993; 74: 763–768.

153. Zhou JA, McIndoe A, Davies H, Sun XY, Crawford L. The induction of cytotoxic T-lymphocyte precursor cells by recombinant vaccinia virus expressing human papillomavirus type 16 L1. Virology 1991; 181: 203–210.

Human Papillomaviruses
D.J. McCance (editor)

Summary—What we need to learn

Dennis J. McCance
Department of Microbiology & Immunology and the Cancer Center, University of Rochester, 601 Elmwood Avenue, Rochester, New York 14642, USA

Human papillomaviruses cause a variety of important epithelial lesions, which at the most severe end of the spectrum lead to life threatening malignant disease. A number of large epidemiological studies over the last decade leave no doubt that the viruses are the causative agent of the various pre-malignant and malignant diseases [1]. However, while infection with certain HPV viruses is essential for premalignant disease, other cellular changes are required for subsequent malignant disease. We know very little of these cellular events and how they determine the fate of the infected cell. Unlike rodent cells, human cells require a number of cellular changes before they exhibit a transformed phenotype. In the case of an experimental model using primary human fibroblasts, kidney or breast epithelial cells, it was shown [2,3] that successful transformation in culture and tumorigenicity in animals required the early region of SV40, both small and large T, plus an oncogenic *ras* (H-*ras*) and the catalytic subunit of the human telomerase gene (hTERT). Therefore, a number of events have to come together for transformation of human cells. The E6 and E7 genes of oncogenic HPV types have combined functions, which are similar, but not identical, to that of SV40 large T. The E6 and E7 proteins from oncogenic viruses efficiently immortalize primary human keratinocytes, which is probably an important function for eventual malignant disease, although these cells are not transformed and are not tumorigenic [4,5]. However, the various functions of each of these viral proteins required for immortalization are not clear. Surprisingly perhaps, the binding of E7 to Rb does not appear to be required for immortalization, although it may be necessary for disease *in vivo*. The fact that mutated *ras* is an important player in transformation of human fibroblasts, kidney and breast epithelial cells and that E7 can cooperate with activated *ras* to transform rodent cells, suggests that the modulation of the Ras/Raf pathway in keratinocytes may be one of the cellular pathways that need to be corrupted for malignant disease to progress. Therefore, we need to know the functions of E6 and E7 that are required along with the combination of cellular changes, to produce a malignant phenotype. In addition, some functions of E4 and E5 [6–8] are consistent with a role in stimulating cell proliferation and their role in HPV pathogenesis needs to be delineated.

An important aspect of HPV biology is that some viral types cause benign disease, while others are associated with malignant disease. However, from the point of view of the virus, they all have to achieve the same goal. All papillomaviruses have to stimulate a differentiating keratinocyte into S-phase so that there are sufficient supplies of the replicative machinery for viral DNA replication. Therefore, one would have thought that common pathways would be used by both groups of viruses. However, almost all of the biological properties exhibited by the high-risk viruses in tissue culture are not found with low-risk types. For instance, HPV 16 can immortalize human keratinocytes, inhibit terminal differentiation of keratinocytes, transform primary rodent fibroblasts in combination with an activated *ras*, and bypass cell cycle arrest signals induced by DNA damage, serum deprivation, TGF-β, and over expression of c-Raf. HPV-6 does not exhibit any of these properties. So there is little biological read out from the benign viruses, leading to the obvious question of how they stimulate G1 to S-phase progression and subsequently replicate.

One interesting biological similarity between the high and the low risk viruses is that both groups of viruses can maintain episomal viral DNA replication in cycling human keratinocytes [6,9]. Long-term replication of the low risk types is curtailed because these viruses cannot immortalize keratinocytes and so the cells senesce. However, during differentiation of the keratinocytes viral DNA amplification was observed for both the high and low types. Therefore, these *in vitro* assays recapitulate the situation in the epithelium and the benign viruses were able to amplify their genomes in arrested cells. Using proteomics, it may be possible to determine the proteins that are regulated by HPV genes during replication in keratinocytes by the two types of virus to determine common modes of action. This may bode well for antiviral research since the E1 and E2 of the high and low risk types can be substituted for each other in replication assays.

While anti-viral research has not been that productive, other ways of tackling the disease by preventing the infection by vaccination have been gaining prominence in the last few years [10]. At present it is unclear what constitutes an effective immune response against HPV, although in regressing warts a lymphocytic infiltrate is observed and cytotoxic T-cells are thought to play a role. An important finding in animal studies was the fact that protective antibody responses are directed against conformational epitopes on the viral capsid and that the L1 protein when expressed in yeast or insect cells can fold in a manner similar to the viral capsid and be recognized by naturally acquired antibodies. Therefore, antibodies may be important in protection against infection. These L1 capsids are called viral like particles (VLPs) and Phase I trials have been carried out in an initial attempt to use VLPs as vaccine against HPV infection or disease. Since antibody responses appear to be type specific it will be necessary to vaccine with multiple types. However, some evidence suggests that T-cell responses may be more cross-reactive. These findings are preliminary and Phase III trials are now in progress in large numbers of individuals to determine if the vaccines protect against infection and disease. There is a strategic difference between infection and disease, since protection against the latter does not necessarily mean protection against infection. Future trials aimed at determining protection against disease will take a number of years (five at least) and many volunteers to determine how protective the vaccine will be.

The activities of HPV proteins, especially E6 and E7, have helped us to understand more about the cell and the pathways involved in cell cycle progression and protein degradation. Future research will no doubt increase this knowledge and possibly identify new targets for either anti-virals or anti-cancer agents or both.

References

1. Wheeler CM. Clinical aspects and epidemiology of HPV infections. In: Human Papillomaviruses (DJ McCance, Editor). Elsevier, Amsterdam, 2002, pp. 1–29.
2. Hahn WC, Counter CM, Lundberg AS, Beijersbergen RL, Brooks MW, Weinberg RA. Creation of human tumour cells with defined genetic elements. Nature 1999; 400: 464–468.
3. Hahn WC, Dessain SK, Brooks MW, King JE, Elenbaas B, Sabatini DM, DeCaprio JA, Weinberg RA, Counter CM, Lundberg AS, Beijersbergen RL. Enumeration of the simian virus 40 early region elements necessary for human cell transformation creation of human tumour cells with defined genetic elements Mol Cell Biol 2002; 22: 2111–2123.
4. Thomas M, Pim D, Banks L. Human papillomavirus E6 protein interactions. In: Human Papillomaviruses (DJ McCance, Editor). Elsevier, Amsterdam, 2002, pp. 71–99.
5. McCance DJ. The Biology of E7. In: Human Papillomaviruses (DJ McCance, Editor). Elsevier, Amsterdam, 2002, pp. 101–118.
6. del Mar Peña L, Laimins LA. Regulation of human papillomavirus gene expression in the vegetative life cycle. In: Human Papillomaviruses (DJ McCance, Editor). Elsevier, Amsterdam, 2002, pp. 31–51.
7. Roberts S. Biology of the E4 protein. In: Human Papillomaviruses (DJ McCance, Editor). Elsevier, Amsterdam, 2002, pp. 119–142.
8. Venuti A, Campo MS. The E5 protein of papillomaviruses. In: Human Papillomaviruses (DJ McCance, Editor). Elsevier, Amsterdam, 2002, pp. 132–164.
9. Liu J-S, Melendy T. Human papillomavirus DNA replication. In: Human Papillomaviruses (DJ McCance, Editor). Elsevier, Amsterdam, 2002, pp. 53–70.
10. Rose R. Human papillomavirus immunology and vaccine development. In: Human Papillomaviruses (DJ McCance, Editor). Elsevier, Amsterdam, 2002, pp. 165–187.

List of Addresses

Lawrence Banks
International Center for Genetic Engineering and Biotechnology
Padriciano 99
I-34012 Trieste, Italy

M. Saveria Campo
Institute of Comparative Medicine
Department of Veterinary Pathology
Glasgow University Veterinary School
Garscube Estate
Glasgow G61 1QH, UK

Loren del Mar Pena
Department of Microbiology–Immunology
Northwestern University Medical School
303 East Chicago
Chicago, IL 60611, USA

Laimonis A. Laimins
Department of Microbiology–Immunology
Northwestern University Medical School
303 East Chicago
Chicago, IL 60611, USA

Jen-Sing Liu
Department of Microbiology
University of Buffalo, State University of New York
School of Medicine and Dentistry and Biomedical Sciences
213 Biomedical Research Building
3435 Main Street
Buffalo, NY 14214, USA

Dennis J. McCance
Department of Microbiology and Immunology and the Cancer Center
University of Rochester, Box 672
610 Elmwood Avenue
Rochester, NY 14642, USA

Thomas Melendy
Department of Microbiology
University of Buffalo, State University of New York
School of Medicine and Dentistry and Biomedical Sciences
213 Biomedical Research Building
3435 Main Street
Buffalo, NY 14214, USA

David Pim
International Center for Genetic Engineering and Biotechnology
Padriciano 99
I-34012 Trieste, Italy

Sally Roberts
Cancer Research UK Institute for Cancer Studies
The Medical School
University of Birmingham
Birmingham B15 2TA, UK

Robert Rose
Departments of Medicine and Microbiology and Immunology
University of Rochester, Box 689
601 Elmwood Avenue
Rochester, NY 14642, USA

Miranda Thomas
International Center for Genetic Engineering and Biotechnology
Padriciano 99
I-34012 Trieste, Italy

Aldo Venuti
Laboratory of Virology
Regina Elena Cancer Institute
Viale Regina Elena, 291
Rome 00161, Italy

Cosette Wheeler
Department of Molecular Genetics and Microbiology
University of New Mexico Health Sciences Center
Albuquerque, NM 87131, USA